Blockchain-enabled Fog and Edge Computing

T0133755

Blockchain-enabled Fog and Edge Computing

Concepts, Architectures, and Applications

Edited by
Dr. Muhammad Maaz Rehan
and
Dr. Mubashir Husain Rehmani

CRC Press
Taylor & Francis Group
Boca Raton London New York

CRC Press is an imprint of the
Taylor & Francis Group, an **informa** business

First edition published 2021
by CRC Press
6000 Broken Sound Parkway NW, Suite 300, Boca Raton, FL 33487-2742

and by CRC Press
2 Park Square, Milton Park, Abingdon, Oxon, OX14 4RN

© 2021 Taylor & Francis Group, LLC

CRC Press is an imprint of Taylor & Francis Group, LLC

ISBN: 978-0-367-45735-8 (hbk)
ISBN: 978-0-367-50744-2 (pbk)
ISBN: 978-1-003-03408-7 (ebk)

Typeset in Sabon
by Lumina Datamatics Limited

Contents

Preface

Blockchain consists of a decentralized ledger of every single transaction that takes place in a peer-to-peer (P2P) network formed by the participants. Blockchain can be permissioned (public), permission-less (private), or hybrid. If blockchain is public, the validator of the transaction can be any participant. If the blockchain is private, the validator of the transaction has to be authorized by the blockchain owner. This blockchain owner can be a single entity validating the transactions and approving the blocks to be added in the blockchain, or it can be a group of entities resulting in the form of consortium blockchain. Potential blockchain applications include electronic voting systems, fund transfers, production, supply chain management, the pharmaceutical industry, the financial sector, and many others.

A transaction flow in blockchain operates in the following manner. If participant 'A' requests some transaction, it is broadcast to the associated P2P blockchain network consisting of computing machines. The transaction is verified by the blockchain participants based on smart contracts, predefined digital rules agreed on by all stakeholders of the system. Once verified, the transaction is considered complete. The best known example of blockchain application is the digital Bitcoin cryptocurrency. The two most popular transaction verification algorithms in the distributed blockchain model for establishing consensus are proof-of-work (PoW) and proof-of-stake (PoS). In PoW, every newly added block has to generate a hash value which should be less than the given target value. This requires many retries and is called 'mining'. Once done, the newly added block is appended to the blockchain, and the hash of previous block is added to it to build the chain. Mining retries make it a time-consuming and energy-intensive process. The Bitcoin cryptocurrency uses PoW verification. Instead of mining, PoS users validate and make changes to the blockchain on the basis of their existing share (called a 'stake') in the currency. A participant with $x\%$ share can validate $x\%$ transactions. This approach provides enormous savings on energy and operating costs. New blockchain systems, like Ethereum, use PoS verification. In short, the blockchain system provides inherent security because of distributed nature of the ledger and verification and validation process of blocks, which makes it extremely difficult to alter transactions.

The term, 'fog computing', was coined by Cisco in January 2014 whose purpose is to make the computing available inside the network instead of sending data to the cloud for computing and then getting feedback from cloud. According to Cisco, fog is the standard, and edge is the concept. A fog device is equipped with computing, storage, and network connectivity resources, and when it is placed at the edge, the processing or analysis of real-time data is performed closest to the nodes which generate data. Thus, switches, routers, embedded servers, video surveillance cameras, and industrial controllers can be fog nodes. Fog computing is needed when decision-making and suggested action is required in milliseconds for situations when thousands or millions of devices across a large geographic area are generating data. Some applications of fog computing are: (i) locking a door, (ii) applying train or car brakes, (iii) zooming a video surveillance camera, (iv) opening a valve if a pressure reading has reached a threshold, and (v) alerting the technician to make preventive repair. As a result of fog computing, decision-making in sensitive situations becomes fast; sensitive traffic remains inside the network premises for processing and decision-making; and less-sensitive information can be sent to the cloud for big data analytics and feedback in the form of up-to-date strategy rules for fog nodes and offloading of huge volume of unnecessary traffic from core network to the cloud.

In the emerging paradigm of fog computing, in which in-network computing near the data producers is on the rise, blockchain-enabled transactions on fog nodes seem a promising solution. For example, if a voting system is based on blockchain, then fog and edge nodes would be the polling stations and the data producers would be voters, the polling staff, and the election commission software. In this, and so many other scenarios, consensus methods and smart contracts can be defined by the blockchain owners for transparently recording everything on fog nodes and later save it to the central cloud for long-term use and analysis. However, there are a number of integration issues in this scenario. For example, a government may consider using a blockchain developed by a third party as a threat or vulnerability and develop their own blockchain with proprietary consensus mechanisms and smart contracts. For different requirements and environments (e.g., factories, universities, banking, voting), both consensus and smart contracts would be different.

Considering the aforementioned importance of integration of blockchain with the emerging fog computing paradigm, this book aims to unveil the working relationship of blockchain and fog and edge computing to solve different everyday problems. The book has been designed in such a way that the reader will not only understand blockchain and fog and edge computing but will also understand their coexistence and their collaborative power to solve different problems. We have given special attention to privacy and security issues related with the blockchain-enabled edge computing paradigm. Overall, this book contains nine chapters which are divided into two parts. Part I covers all fundamental concepts of blockchain and the applications of blockchain together with edge computing, and Part II covers the security and privacy issues of blockchain-enabled edge computing.

Part I: Fundamental concepts and applications of blockchain-enabled fog and edge computing

In Chapter 1, authors discuss Ethereum with respect to its capability to develop new models of distributed ownership and Tactile Internet. Recent progress on Ethereum and open challenges are discussed in detail. The chapter also discusses similarities and differences between Bitcoin-based blockchains and Ethereum-based blockchains. The comparison determines that Ethereum is a better choice as a blockchain platform. The authors also argue that Ethereum is better than other blockchains because of a number of salient features including its cooperative and friendly working with artificial intelligence and robots and also that it is highly compatible with decentralized edge computing solutions. As a result, Ethereum-based solutions would birth a new hybrid form of collaboration in which intelligence and automation are at the centre, and humans would be at the edges. The authors conclude the chapter with open research challenges and future work.

Chapter 2 is a review article in which the authors discuss the interworking of the Internet of Things (IoT) and 5G in the presence of blockchain technologies keeping in view the top-tier Smart city environment, New York City. The authors offer some pragmatic views of different evolving disciplines based on several projects in which they are involved. The survey discusses in detail the requirements of IoT with edge structures; 5G cellular connectivity in urban and suburban environments, cyber-security, and blockchain; Smart city IoT applications with and without blockchains and edge nodes; and the applications of all of these in the realm of health, supply chain, surveillance and 'ring of steel', insurance, and banking. At the end, the authors state future research directions.

In Chapter 3, the authors focus on the Internet of Vehicles (IoV) environment and present a blockchain-based solution for traffic challan resulting from speeding in the absence of speed cameras and traffic wardens. Vehicles use permissioned blockchains on highways and elect the miner node. Offending vehicles are detected by neighbours and then the offending event transaction is validated by all neighbours that noticed the offense. The transaction is added to the block by the miner. Once, the miner reaches the edge node, a block offloading point at the petrol pump or rest area, it offloads the block which is added to the blockchain of that road.

In Chapter 4, the authors focus impersonation and tempering attacks on IoT devices and provide a hardware-based solution for them in the presence of blockchain and edge computing. The authors argue that because IoT devices have everything limited, these devices can employ a state-of-art security mechanism. As a result, IoT devices are more prone to hacking. The hardware-primitive-based blockchain is the suggested solution. The authors use physical unclonable functions (PUFs) with blockchain and smart contracts that provide data integrity and safeguard IoT devices from data tampering.

Part II: Security and privacy issues in blockchain-enabled fog and edge computing

In Chapter 5, the authors argue that in the presence of a large number of IoT devices, providing efficient decision-making with the least response time as well as data protection is a great challenge. When blockchain works in the presence of fog computing, both privacy and security have not been addressed. The authors, therefore, present a three-layer architecture that unites the strengths of blockchain and fog computing and solves the afore-mentioned challenge.

In Chapter 6, authors review the coexistence of blockchain with fog and edge architecture and outline their importance in coping with different security attacks. The main focus of the chapter is to draw the reader's attention towards the contribution of blockchain features within a fog and edge environment when handling denial-of-service, distributed denial-of-service (DDoS), man-in-the-middle, and other such attacks. The authors also highlight the significance of edge node in preserving the privacy of network knowledge and outline the challenges associated with the reference architecture that combines blockchain with the fog and edge environment. The authors conclude the chapter by highlighting resource management issues while integrating the two technologies and presenting future research directions.

In Chapter 7, the authors argue that different security challenges are actually faced at the edge with the use of conventional cloud services. Over the edge, blockchain can be used to overcome the most crucial challenge of security, scalability, and processing power of devices. Because the distributed copy of blockchain locally validates any existing device of the system, using blockchain at the network edge does not need a persistent connectivity with the network. The authors favour edge-based blockchain solutions in which devices have an associated profile along with a public key in it, and all devices are stored in the blockchain network along with their profiles. Thus, without relying on a third party or connecting to a cloud service, local verification of devices can be done through the profiles stored in the blockchain. Because edge nodes authenticate IoT devices and initiate secure connections, content integrity is also preserved. Although blockchain provides security over the edge, it increases resource utilization along with high computational overhead and storage.

In Chapter 8, the authors cover security, privacy, and data integrity aspects that occur in the presence of the blockchain-enabled fog environment. In diverse data exchange, sensitive or confidential information may be present which would require a high level of security. The authors not only cover the advantages which are offered by the IoT, fog, and blockchain alliance but also investigate the scalability issue. Finally, the authors conclude by presenting some solutions that help achieve secure storage when blockchain is deployed in a fog and edge environment.

Finally, in Chapter 9, authors present a privacy-preserving mechanism for smart grids. They use the advantageous characteristics of blockchain, edge computing, and differential privacy on a specific scenario of a smart grid, 'on-demand readings'. The authors then develop a blockchain based differentially on a private on-demand edge reporting (BDOR) approach in which smart meter users can efficiently share their data without worrying about their privacy and lack of computational power.

This book is written for students, computer scientists, researchers, and developers who wish to work in the domain of blockchain and fog computing. One of the unique features of this book is highlighting the issues, challenges, and future research directions associated with the blockchain-enabled fog computing paradigm. We hope the readers will enjoy reading this book!

Muhammad Maaz Rehan and Mubashir Husain Rehmani

Acknowledgements

We would like to express sincere thanks to ALLAH Subhanahu Wa-ta'ala that by His grace and bounty we were able to edit this book. We dedicate this work to the last Prophet, Muhammad (S.A.W.W.).

Maaz wants to acknowledge his parents, wife, and kids (Aamnah Murvareed, Muhammad Bin Maaz, and Ahmad Bin Maaz) for their prayers and support throughout.

Mubashir wishes to express gratitude to Sheikh, Mufti Mohammad Naeem Memon of Hyderabad, Pakistan. He could not have edited this book without his prayers, spiritual guidance, and moral support. He also wants to acknowledge his family for their continued support.

Muhammad Maaz Rehan and Mubashir Husain Rehmani

Editors

Muhammad Maaz Rehan (M'16, SM'17) is an Assistant Professor in COMSATS University Islamabad, Wah Campus, Pakistan, under the Tenure Track System and has been attached with academia and industry for the last 15 years. He leads the Telecom and Networks (TelNet) Research Group (https://sites.google.com/view/telnet-rg/) at COMSATS University. He obtained a PhD from Universiti Teknologi PETRONAS (UTP), Malaysia, in 2016 with two Bronze medals for his PhD work. He is an Editor of the IEEE Softwarization Newsletter and Associate Editor of IEEE Access & Springer Human-centric Computing and Information Sciences journals. Dr. Maaz has been a fellow of Internet Society (ISOC) for the Internet Engineering Task Force (IETF) twice. He is lead author of the book, *Blockchain-enabled Fog and Edge Computing: Concepts, Architectures and Applications*. He has supervised 13 students in master's degrees at COMSATS and two students abroad in dissertation work. The research areas of Dr. Maaz's research interests include blockchain, Internet of Things and Internet of Vehicles, ICN, machine learning for networking, and fog and edge computing.

Mubashir Husain Rehmani (M'14-SM'15) received a B.Eng. in computer systems engineering from Mehran University of Engineering and Technology, Jamshoro, Pakistan, in 2004, an MS from the University of Paris XI, Paris, France, in 2008, and a PhD from the University Pierre and Marie Curie, Paris, in 2011. He is currently working as Assistant Lecturer in the Department of Computer Science, Cork Institute of Technology, Ireland. Prior to this, he worked as postdoctoral researcher at the Telecommunications Software and Systems Group (TSSG), Waterford Institute of Technology (WIT), Waterford, Ireland. He also served as an Assistant Professor at COMSATS Institute of Information Technology, Wah Cantt, Pakistan, for 5 years. He is currently an Area Editor of the *IEEE Communications Surveys and Tutorial* and served as an Associate Editor of the *IEEE Communications Surveys and Tutorials* from 2015 to 2017. He also serves as a Column Editor for Book Reviews in *IEEE Communications Magazine* and as Associate Editor of *IEEE Communications Magazine*, Elsevier *Journal of Network and Computer Applications (JNCA)* and the *Journal of Communications and Networks (JCN)*. He also serves as a Guest Editor of the Elsevier *Ad Hoc Networks* journal, the Elsevier *Future Generation Computer Systems* journal, the *IEEE Transactions on Industrial Informatics*, and the Elsevier *Pervasive and Mobile Computing* journal. He has authored or edited two books published by IGI Global; one book published by CRC Press; and one book published by Wiley, UK. He received 'Best Researcher of the Year 2015 of COMSATS Wah' award in 2015 and received the certificate of appreciation, 'Exemplary Editor of the IEEE Communications Surveys and Tutorials for the year 2015' from the IEEE Communications Society. He received Best Paper Award from IEEE ComSoc Technical Committee on Communications Systems Integration and Modeling (CSIM) in IEEE ICC 2017. He consecutively received research productivity awards in 2016–2017 and also ranked #1 in all Engineering disciplines from Pakistan Council for Science and Technology (PCST), Government of Pakistan. He received Best Paper Award in 2017 from Higher Education Commission (HEC), Government of Pakistan. He is the recipient of Best Paper Award in 2018 from Elsevier *Journal of Network and Computer Applications*.

Contributors

Ali Nawaz Abbasi

Moayad Aloqaily
Al Ain University
UAE and xAnalytics Inc
Ottawa, Ontario, Canada

Muhammad Naveed Aman
Department of Computer
    Science
National University of Singapore
Singapore

Imane Ameli
University Moulay Ismail
Meknes, Morocco

Tashjia Anfal

Samiha Ayed
Institut Charles Delaunay
Team ERA Université de
    Technologie de Troyes
Troyes, France

Nabil Benamar
University Moulay Ismail
Meknes, Morocco

Abdeljalil Beniiche
Optical Zeitgeist Laboratory
Centre Énergie, Matériaux et
    Télécommunications
Institut National de la Recherche
    Scientifique (INRS)
Montrtut, Quebec, Canada

Ouns Bouachir
Zayed University, UAE

Jinjun Chen
Swinburne University of
    Technology
Hawthorn, Victoria, Australia

Amin Ebrahimzadeh
Optical Zeitgeist Laboratory
Centre Énergie, Matériaux et
    Télécommunications
Institut National de la Recherche
    Scientifique (INRS)
Montréal, Quebec, Canada

Moez Esseghir
Institut Charles Delaunay
Team ERA Université de
    Technologie de Troyes
Troyes, France

Rima Grati
Zayed University, UAE

Abdelhakim Senhaji Hafid
University of Montreal
Montréal, Quebec, Canada

Muneeb Ul Hassan
Swinburne University of
    Technology
Hawthorn, Victoria,
    Australia

Uzair Javaid
Department of Electrical and
    Computer Engineering

Alejandro Jurnet
Advanced Network Architectures
    Lab, CRAAX
Universitat Politècnica de
    Catalunya, UPC
Barcelona, Spain

Chaima Khalfaoui
Institut Charles Delaunay
Team ERA Université de
    Technologie de Troyes
Troyes, France

Martin Maier
Optical Zeitgeist Laboratory
Centre Énergie, Matériaux et
    Télécommunications
Institut National de la
    Recherche Scientifique
    (INRS)
Montréal, Quebec, Canada

Pau Marcer
Advanced Network Architectures
    Lab, CRAAX
Universitat Politècnica de
    Catalunya, UPC
Barcelona, Spain

Eva Marin
Advanced Network Architectures
    Lab, CRAAX
Universitat Politècnica de
    Catalunya, UPC
Barcelona, Spain

Xavier Masip
Advanced Network Architectures
    Lab, CRAAX
Universitat Politècnica de
    Catalunya, UPC
Barcelona, Spain

Daniel Minoli

Adel Ben Mnaouer
Canadian University Dubai, UAE

Benedict Occhiogrosso
DVI Communications Inc.
New York

Muhammad Maaz Rehan

Mubashir Husain Rehmani
Department of Computer Science
Cork Institute of Technology
Cork, Ireland

Biplab Sikdar
Department of Electrical and
    Computer Engineering

Part I

Fundamental concepts and applications of blockchain-enabled fog and edge computing

Chapter 1

From blockchain Internet of Things (B-IoT) towards decentralising the Tactile Internet

Abdeljalil Beniiche, Amin Ebrahimzadeh, and Martin Maier

CONTENTS

1.1 INTRODUCTION

The Internet has been constantly evolving from the mobile Internet to the emerging Internet of Things (IoT) and future Tactile Internet. Similarly, the capabilities of future 5G networks will extend far beyond those of previous generations of mobile communication. Beside 1000-fold gains in area capacity, 10 Gb/s peak data rates, and connections for at least 100 billion devices, an important aspect of the 5G vision is *decentralisation*. While 2G, 3G, and 4G cellular networks were built under the design premise of having complete control at the infrastructure side, 5G systems may drop this design assumption and evolve the cell-centric architecture into a more device-centric one. Although there is a significant overlap of design objectives among 5G, IoT, and the Tactile Internet – most notably, ultra-reliable and low-latency communication (URLLC) – each one of them exhibits unique characteristics in terms of underlying communications paradigms and enabling end devices [1].

Today's Internet is ushering in a new era. While the first generation of digital revolution brought us the Internet of information, the second generation – powered by decentralised blockchain technology – is bringing us the Internet of value, a true peer-to-peer (P2P) platform that has the potential to go far beyond digital currencies and record virtually everything of value to humankind in a distributed fashion without powerful intermediaries [2]. Some refer to decentralised blockchain technology as the 'alchemy of the 21st century' because it may leverage end-user equipment for converting computing into digital gold. More importantly, though, according to Don and Alex Tapscott, the blockchain technology enables trusted collaboration that can start to change the way wealth is distributed because people can share more fully in the wealth they create, rather than trying to solve the problem of growing social inequality through the redistribution of wealth only. As a result, decentralised blockchain technology helps create platforms for distributed capitalism and a more inclusive economy, which works best when it works for everyone as the foundation for prosperity. Furthermore, the authors of [3] pointed out the important role of blockchain and distributed ledger technology (DLT) applications as a next generation of distributed sensing services for 6G-driving applications whose need for connectivity will require a synergistic mix of URLLC and massive machine type communications (mMTC) to guarantee low latency, reliable connectivity, and scalability. Furthermore, blockchains and smart contracts can improve the security of a wide range of businesses by ensuring that data cannot be damaged, stolen, or lost. In [4], the authors presented a comprehensive survey on the use of blockchain technologies to provide distributed security services. These services include entity authentication, confidentiality, privacy, provenance, and integrity assurances.

A blockchain technology of particular interest is Ethereum, which went live in July 2015. Ethereum made great strides in having its technology accepted as the blockchain standard, when Microsoft Azure started offering it as a service in November 2015.[1] Ethereum was founded by Canadian Vitalik Buterin after his request for creating a wider and more general scripting language for the development of decentralised applications (DApps) that are not limited to cryptocurrencies, a capability that Bitcoin lacked, was rejected by the Bitcoin community [5]. Ethereum enables new forms of economic organisation and distributed models of companies, businesses, and ownership (e.g., self-organised holacracies and member-owned cooperatives). Or as Buterin puts it, while most technologies tend to automate workers on the periphery doing menial tasks, Ethereum automates away the centre. For instance, instead of putting the taxi driver out of a job, Ethereum puts Uber out of a job and lets the taxi drivers work with the customer directly (before Uber's self-driving cars will eventually wipe out their jobs). Hence, Ethereum does not aim at eliminating jobs, so much as it changes the definition of work. In fact, it gave rise to the first decentralised autonomous organisation (DAO) built within the Ethereum project. The DAO is an open-source, distributed software that exists 'simultaneously nowhere and everywhere', thereby creating a paradigm shift that offers new opportunities to democratise business and enables entrepreneurs of the future to design their own virtual organisations customised to the optimal needs of their mission, vision, and strategy to change the world [6].

There exist excellent surveys on Bitcoin and other decentralised digital currencies (e.g., [7]). Likewise, the fundamental concepts and potential of blockchain technologies for society and industry in general have been described comprehensively in various existent tutorials (e.g., [8]). In this chapter, we focus on how blockchain technologies can be used in an IoT context by providing an up-to-date survey on recent progress and open challenges for realising the emerging blockchain IoT (B-IoT). Unlike the IoT without any human involvement in its underlying machine-to-machine (M2M) communications, the Tactile Internet is anticipated to keep the human in (rather than out of) the loop by providing real-time transmission of haptic information, (i.e., touch and actuation) for the remote control of physical or virtual objects through the Internet. Towards this end, we elaborate on how Ethereum blockchain technologies, in particular the DAO, may be leveraged to realise future techno-social systems, notably the Tactile Internet, which is yet unclear in how exactly it would work [8].

The remainder of this chapter is structured as follows. In Section 1.2.1, we first explain the commonalities of and specific differences between Ethereum and Bitcoin blockchains in greater detail. Section 1.3.1 then reviews recent progress and open challenges of the emerging B-IoT. In Section 1.4, after briefly reviewing the key concepts of the emerging Tactile Internet, we introduce the so-called human-agent-robot teamwork (HART) design approach and our proposed low-latency FiWi enhanced LTE-A HetNets based on

---

[1] M. Gray, 'Ethereum Blockchain as a Service now on Azure,' https://azure.microsoft.com.

advanced multi-access edge computing with embedded artificial intelligence (AI) capabilities. In Section 1.5, we elaborate on the potential role of Ethereum and, in particular the DAO, in helping decentralise the Tactile Internet. Section 1.6 discusses the symbiosis of blockchain, AI, and augmented intelligence in more detail, and Section 1.7 suggests future research areas. Finally, Section 1.8 concludes the chapter.

1.2 BLOCKCHAIN TECHNOLOGIES

In this section, we give a brief overview of the basic concepts of blockchain technologies, paying a particular attention to the main commonalities and specific differences between Ethereum and Bitcoin. We then introduce the DAO, which represents a salient feature of Ethereum that cannot be found in Bitcoin.

1.2.1 Ethereum versus Bitcoin blockchains

Blockchain technologies have been undergoing several iterations as both public organisations and private corporations seek to take advantage of their potential. A typical blockchain network is essentially a distributed database (also known as a 'ledger'), comprising records of all transactions or digital events that have been executed by or shared among participating parties. Blockchains may be categorised into public (i.e., permissionless) and private (i.e., permissioned) networks. In the former category, anyone may join and participate in the blockchain. Conversely, a private blockchain applies certain access control mechanisms to determine who can join the network. A public blockchain is immutable because none of the transactions can be tampered with or changed. Also, it is pseudo-anonymous because the identity of those involved in a transaction is represented by an address key in the form of a random string. Table 1.1 highlights the major differences between public and private blockchains, as will be discussed in further detail. Note that both Ethereum and Bitcoin are public blockchains.

Figure 1.1 illustrates the main commonalities of and differences between Bitcoin and Ethereum blockchains. The Bitcoin blockchain is predominantly

Table 1.1 Public versus private blockchains

	Public blockchain	*Private blockchain*
Network Type	Fully decentralised	Partially decentralised
Access	Permissionless read/write	Permissioned read/write
User Identity	Pseudo-anonymous	Known participants
Consensus Mechanism	Proof-of-work/Proof-of-state	Pre-approved participants
Consensus Determination	By all miners	By one organisation
Immutability	Nearly impossible to tamper	Could be tampered
Purpose	Any decentralised applications	Business applications

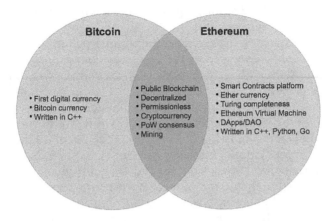

Figure 1.1 Bitcoin and Ethereum blockchains: Commonalities and differences.

designed to facilitate Bitcoin transactions. It is the world's first fully functional digital currency that is truly decentralised, open source, and censorship resistant. Bitcoin makes use of a cryptographic proof-of-work (PoW) consensus mechanism based on the SHA-256 hash function and digital signatures. Achieving consensus provides extreme levels of fault tolerance, ensures zero downtime, and makes data stored on the blockchain forever unchangeable and censorship-resistant in that everyone can see the blockchain history, including any data or messages. There are two different types of actors, whose roles are defined as follows:

- *Regular nodes*: A regular node is a conventional actor, who just has a copy of the blockchain and uses the blockchain network to send or receive Bitcoins.
- *Miners*: A miner is an actor with a particular role, who builds the blockchain through the validation of transactions by creating blocks and submitting them to the blockchain network to be included as blocks. Miners serve as protectors of the network and can operate from anywhere in the world as long as they have sufficient knowledge about the mining process, the hardware and software required to do so, and an Internet connection.

In the Bitcoin blockchain, a block is mined about every 10 minutes, and the block size is limited to 1 MByte. Note that the Bitcoin blockchain is restricted to a rate of seven transactions per second, which renders it unsuitable for high-frequency trading. Other weaknesses of the Bitcoin blockchain include its script language, which offers only a limited number of small instructions and is non-Turing-complete. Furthermore, developing applications using the Bitcoin script language requires advanced skills in programming and cryptography.

Ethereum is currently the second-most popular public blockchain after Bitcoin. It has been developed by the Ethereum Foundation, a Swiss non-profit organisation, with contributions from all over the world. Ethereum has its own cryptocurrency called 'Ether', which provides the primary form of liquidity allowing for exchange of value across the network. Ether also provides the mechanism for paying and earning transaction fees that arise from supporting and using the network. Like Bitcoin, Ether has been the subject of speculation witnessing wide fluctuations. Ethereum is well suited for developing DApps that need to be built quickly and interact efficiently and securely via the blockchain platform. Similar to Bitcoin, Ethereum uses a PoW consensus method for authenticating transactions and proving the achievement of a certain amount of work. The hashing algorithm used by the PoW mechanism is called 'Ethash'. Different from Bitcoin, Ethereum developers expect to replace PoW with a so-called 'proof-of-stake' (PoS) consensus. PoS will require Ether miners to hold some amount of Ether, which will be forfeit if the miner attempts to attack the blockchain network. The Ethereum platform is often referred to as a 'Turing-complete Ethereum virtual machine (EVM)' built on top of the underlying blockchain. Turing-completeness means that any system or programming language is able to compute anything computable, provided it has enough resources. Note that the EVM requires a small amount of fees for executing transactions. These fees are called 'gas', and the required amount of gas depends on the size of a given instruction. The longer the instruction, the more gas is required.

Whereas the Bitcoin blockchain simply contains a list of transactions, Ethereum's basic unit is the 'account'. The Ethereum blockchain tracks the state of every account, whereby all state transitions are transfers of value and information between accounts. The account concept is considered an essential component and data model of the Ethereum blockchain because it is vital for a user to interact with the Ethereum network via transactions. Accounts represent the identities of external agents (e.g., human or automated agents, mining nodes). Accounts use public key cryptography to sign each transaction such that the EVM can securely validate the identity of the sender of the transaction.

Beside C++, Ethereum supports several programming languages based on JavaScript and Python (e.g., Solidity, Serpent, Mutan, or LLL), whereby Solidity is the most popular language for writing so-called 'smart contracts'. A smart contract is an agreement that runs exactly as programmed without any third-party interference. It uses its own arbitrary rules of ownership, transaction formats, and state-transition logic. Each method of a smart contract can be invoked via either a transaction or another method. Smart contracts enable the realisation of

DApps, which may look exactly the same as conventional applications with regard to application programming interface (API), though the centralised backend services are replaced with smart contracts running on the decentralised Ethereum network without relying on any central servers. Interesting examples of existent DApps include Augur (a decentralised prediction market), Weifund (an open platform for crowdfunding), Golem (supercomputing), and Ethlance (decentralised job market platform), among others. To provide an effective means of communications between DApps, Ethereum uses the Whisper P2P protocol, a fully decentralised middleware for secret messaging and digital cryptography. Whisper supports the creation of confidential communication routes without the need for a trusted third party. It builds on a peer sampling service that takes into account network limitations such as network address translation (NAT) and firewalls. In general, any centralised service may be converted into a DApp by using the Ethereum blockchain.

1.2.2 Decentralised autonomous organisations (DAOs)

The most remarkable thing about cryptocurrencies and blockchain might be how they enable people and organisations on a global level, all acting in their own interest, to create something of immense shared value. Many observers assert that this is a real alternative to current companies. The decentralisation, crowd-based technologies of cryptocurrencies, distributed ledgers, distributed consensus, and smart contracts provide the possibility to fundamentally change the way people organise their affairs and offer a new paradigm for enterprise design. Two recent efforts to substitute a crowd for a company are blockchain technology and DAOs. DAOs are decentralised organisations without a central authority or leader. They operate on a programming code that is encoded on the Ethereum blockchain. Like the blockchain, the code of a DAO moves away from traditional organisations by removing the need for centralised control. Not even the original developers of the DAO have any extra authority because it runs independently without any human intervention. It may be funded by a group of individuals who cover its basic costs and give the funders voting rights rather than any kind of ownership or equity shares. This creates an autonomous and transparent system that will continue on the network for as long as it provides a useful service to its customers.

A successful example of deploying the DAO concept for automated smart contract operation is Storj, which is a decentralised, secure, private, and encrypted cloud storage platform that may be used as an alternative to centralised storage providers like Dropbox or Google Drive. A DAO may be funded by a group of individuals who cover its basic costs, giving the funders voting rights rather than any kind of ownership or equity shares. This creates an autonomous and transparent system that will continue on the network for as long as it provides a useful service for its customers. DAOs exist as open-source, distributed software that executes smart contracts and works according to specified governance rules and guidelines. Buterin described the ideal of a DAO on the Ethereum Blog as follows: 'It is an entity that lives on the Internet and exists autonomously, but also heavily relies on hiring individuals to perform certain tasks that the automation itself cannot do.' Unlike AI-based agents that are completely autonomous, a DAO still requires heavy involvement from humans specifically interacting according to a protocol defined by the DAO to operate. For illustrating the distinction between a DAO and AI, Figure 1.2 shows a quadrant chart that classifies DAOs, AI, traditional organisations as well as robots, which have been widely deployed in assembly lines among others, with regard to automation and humans involved at their edges and centre. We will elaborate on how this particular feature of DAOs (i.e., automation at the centre and humans at the edges) can be exploited for decentralising the Tactile Internet in Section 1.5. Towards this end, we also briefly note that according to Buterin a DAO is non-profit, though one can make money in a DAO, not by providing investment into the DAO itself but by participating in its ecosystem (e.g., via membership).

For convenience, Table 1.2 summarises the technical details of our comparison of Bitcoin and Ethereum blockchains.

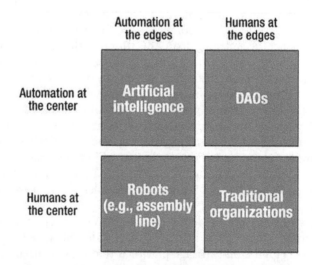

Figure 1.2 DAOs versus artificial intelligence, traditional organisations, and robots (widely deployed in assembly lines, among others): Automation and humans involved at their edges and centre. (From Ethereum Blog).

Table 1.2 Comparison of bitcoin and ethereum blockchains

	Bitcoin	Ethereum
Currency	Bitcoin	Ether
Applications	Cryptocurrency	Cryptocurrency, Smart Contract, DApps/DAO
Written in	C++	C++, Go, Python
Consensus	Proof-of-work (based on SHA-256)	Proof-of-work (Ethash), Planning for proof-of-stake
Turing Completeness	No	Yes
Anonymity Mechanisms	No	Yes (with Whisper protocol)
Censorship Resistance	No	Yes
Transaction Limit	7 transactions/sec	20 transactions/sec
State Concept	No	Data
Smart Contract Languages	No	Solidity, Serpent, Mutan, LLL
Smart Contract Execution	No	Ethereum Virtual Machine (EVM)
Block Time	10 minutes	15 seconds
Data Model	Transaction-based	Account-based
Client P2P Connections	No	Yes (with Whisper protocol)
Routing	No	Yes (Whisper protocol)

1.3 BLOCKCHAIN IoT AND EDGE COMPUTING

In this section, after defining the integration of blockchain and IoT (B-IoT), we discuss the motivation of such integration followed by a description of the challenges of integrating blockchain and edge computing.

1.3.1 Blockchain IoT (B-IoT): Recent progress and related work

Recall from Section 1.1 that the IoT is designed to enable communications among machines without relying on any human involvement. Thus, its underlying M2M communications is useful for enabling the automation of industrial and other machine-centric processes. The emerging B-IoT represents a powerful combination of two massive technologies – blockchain and M2M communications – that allows us to automate complex multistep IoT processes (e.g., via smart contracts). With the ever-increasing variety of communication protocols between IoT devices, there is a need for transparent yet highly secure and reliable IoT device management systems. This section surveys the state of the art of the emerging B-IoT, describing recent progress and open challenges.

The majority of IoT devices are resource constrained, which restricts them to be part of the blockchain network. To cope with these limitations, the author of [9] proposed a decentralised access management system, where all entities are part of an Ethereum blockchain except for IoT devices as well as so-called 'management hub' nodes that request permissions from the blockchain on behalf of the IoT devices belonging to different wireless sensor networks. In addition, entities called 'managers' interact with the smart contract hosted at a specific 'agent node' in the blockchain to define or modify the access control policies for the resources of their associated IoT devices. The proof-of-concept implementation evaluated the new system architecture components that are not part of the Ethereum network (i.e., management hub and IoT devices) and demonstrated the feasibility of the proposed access management architecture in terms of latency and scalability. Another interesting Ethereum case study can be found in [10], which reviews readily available Ethereum blockchain packages for realising a smart home system according to its smart contract features for handling access control policy, data storage, and data flow management.

The architectural issues for realising blockchain-driven IoT services were investigated in greater detail in [11]. In a preliminary study using a smart-thing-renting service as an example B-IoT service, the authors compared the following four different architectural styles based on Ethereum: (i) fully centralised (cloud without blockchain), (ii) pseudo-distributed things (physically located in central cloud), (iii) distributed things (directly controlled by smart contract), and (iv) fully distributed. The preliminary results indicate that a fully distributed architecture, where a blockchain endpoint is deployed on the end-user device, is superior in terms of robustness and security.

The various perspectives for integrating secure elements in Ethereum transactions were discussed in [12]. A novel architecture for establishing trust in Ethereum transactions exchanged by smart things was presented. To prevent the risks that secret keys for signature are stolen or hacked, the author proposed to use javacard-secure elements and a so-called 'crypto currency smart card' (CCSC). Two CCSC use cases were discussed. In the first one, the CCSC was integrated in a low-cost B-IoT device powered by an Arduino processor, in which sensor data are integrated in Ethereum transactions. The second use case involved the deployment of CCSC in remote APDU call secure (RACS) servers to enable remote and safe digital signatures by using the well-known elliptic curve digital signature algorithm (ECDSA).

Blockchain transactions require public-key encryption operations such as digital signatures. However, not all B-IoT devices can support this computationally intensive task. For this reason, in [13], the authors proposed a preliminary design of a gateway-oriented approach, where all blockchain-related operations are offloaded to a gateway. The authors noted that their approach is compatible with the Ethereum client side architecture.

Because of the massive scale and distributed nature of IoT applications and services, blockchain technology can be exploited to provide a secure, tamper-proof B-IoT network. More specifically, the key properties of tamper-resistance and decentralised trust allow us to build a secure authentication and authorisation service, which does not have a single point of failure. Towards this end, the authors of [14] made a preliminary attempt to develop a security model backed by blockchain that provides confidentiality, integrity, and availability of data transmitted and received by nodes in a B-IoT network. The proposed solution encompasses a blockchain protocol layer on top of the TCP/IP transport layer and a blockchain application layer. The first one comprises a distributed consensus algorithm for B-IoT nodes, whereas the latter one defines the IoT security-specific transactions and their semantics for the higher protocol layers. To evaluate the feasibility and performance of the proposed layered architecture, B-IoT nodes connected in a tree topology were simulated using 1 Gbps Ethernet or 54 Mbps WiFi links. The simulation results showed that the block arrival rate was not affected much by the increased latency and reduced bandwidth, when replacing wired Ethernet with wireless WiFi links because the block difficulty adjustment adapts dynamically to the network conditions.

Among various low-power wide-area (LPWA) technologies, long-range (LoRa) wireless radio frequency (RF) is considered one of the most promising enabling technologies for realising massive IoT deployment. In [15], the authors presented a proof-of-concept demonstrator to enable low-power, resource-constrained LoRa IoT end devices to access an Ethereum blockchain network via an intermediate gateway, which acts as a full blockchain node. More specifically, a battery-powered IoT end device sends position data to the LoRa gateway, which in turn forwards it through the standard Go-lang-based Ethereum client Geth to the blockchain network using a smart contract. An event-based communication mechanism between the LoRa gateway and a backend application server was implemented as proof-of-concept demonstrator.

One of the fundamental challenges of object identification in IoT stems from the traditional domain name system (DNS). Typically, DNS is managed in centralised modules and, thus, may cause large-scale failures as a result of unilateral advanced persistent threat (APT) attacks as well as zone file synchronisation delays in larger systems. Clearly, a more robust and distributed name management system is needed that supports the smooth evolution of DNS and renders it more efficient for IoT and the future Internet in general. Towards this end, a decentralised blockchain-based domain name system called 'DNSLedger' was introduced in [16]. To rebuild the hierarchical structure of DNS, DNSLedger contains two kinds of blockchain: (i) a single root chain that stores all the top-level domain information and (ii) multiple top-level domain (TLD) chains, each responsible for the information about its respective domain name. In DNSLedger, servers of domain names act as blockchain nodes, while each TLD chain may select one or more servers to join the root chain. DNSLedger clients may execute common DNS functions such as domain name look up, application, and modification.

Many of the aforementioned studies considered Ethereum as the blockchain of choice. It was shown that fully distributed Ethereum architectures are able to enhance both robustness and security. Furthermore, a gateway-oriented design approach was often applied to offload computationally intensive tasks from low-power, resource-constrained IoT end devices onto an intermediate gateway and, thus, enable them to access the Ethereum blockchain network. Also, it was shown that the block arrival rate does not deteriorate much by the increased latency and reduced bandwidth of WiFi access links.

Despite the recent progress, the salient features that set Ethereum aside from other blockchains (see Section 1.2.1) remain to be explored in more depth, including their symbiosis with other emerging key technologies such as AI and robots as well as decentralised cloud computing solutions known as 'edge computing'.

A question of particular interest hereby is how decentralised blockchain mechanisms may be leveraged to let new hybrid forms of collaboration among individuals, which have not been entertained in the traditional market-oriented economy dominated by firms rather than individuals, emerge [8].

1.3.2 Blockchain-enabled edge computing

One of the critical challenges in cloud computing is the end-to-end responsiveness between the mobile device and an associated cloud. To address this challenge, multi-access edge computing (MEC) is proposed, which is a mobility-enhanced small-scale cloud data centre that is located at the edge of the Internet (e.g., Radio Access Network [RAN]) and in close proximity to mobile subscribers. An MEC entity is a trusted, resource-rich computer or cluster of computers that is well-connected to the Internet and available for use by nearby mobile devices. According to the white paper published by ETSI, MEC is considered a key emerging technology to be an important component of next-generation networks. In light of the aforementioned arguments, the integration of blockchain and edge computing into one unified entity becomes a natural trend. On one hand, by incorporating blockchain into the edge computing network, the system can provide reliable access and control of the network, computation, and storage over decentralised nodes. On the other hand, edge computing enables blockchain storage and mining computation from power-limited devices. Furthermore, off-chain storage and off-chain computation at the edges enable scalable storage and computation on the blockchain [17].

Several recent studies on blockchain and edge computing have been carried out. In [18], resource-constrained IoT devices were released from computation-intensive tasks by offloading blockchain consensus processes and data-processing tasks onto more powerful edge computing resources. The proposed EdgeChain was built on the Ethereum platform and uses smart contracts to monitor and regulate the behaviour of IoT devices based on how they act and use resources. Another blockchain-enabled computation offloading scheme for IoT with edge computing capabilities, called 'BeCome', was proposed in [19]. The authors of this study aimed at decreasing the task offloading time and energy consumption of edge computing devices, while achieving load balancing and data integrity.

The study in [20] proposed a blockchain-based trusted data management scheme called 'BlockTDM' for edge computing to solve the data trust and security problems in an edge computing environment. Specifically, the authors proposed a flexible and configurable blockchain architecture that includes a mutual authentication protocol, flexible consensus, smart contract, block and transaction data management as well as blockchain node management and deployment. The BlockTDM scheme is able to support matrix-based multichannel data segment and isolation for sensitive or privacy data protection. Moreover, the authors designed user-defined sensitive data encryption before the transaction payload is stored in the blockchain system. They implemented a conditional access and decryption query of the protected blockchain data and transactions through an appropriate smart contract. Their analysis and evaluation show that the proposed BlockTDM scheme provides a general, flexible, and configurable blockchain-based paradigm for trusted data management with high credibility.

In summary, blockchain-enabled edge computing has become an important concept that leverages decentralised management and distributed services to meet the security, scalability, and performance requirements of next-generation of communications networks, as discussed in technically greater detail in Section 1.5.

1.4 THE IEEE P1918.1 TACTILE INTERNET

In this section, we give a brief overview of the basic concepts, features, structure, and taxonomy of the Tactile Internet. Subsequently, some of the typical Tactile Internet applications and network infrastructure requirements are presented in greater detail.

1.4.1 The Tactile Internet: Key principles

The term 'Tactile Internet' was first coined by Fettweis in 2014. In his seminal paper [21], the Tactile Internet was defined as a breakthrough enabling unprecedented mobile applications for tactile steering and control of real and virtual objects by requiring a round-trip latency of 1–10 ms. Later in 2014, ITU-T published a Technology Watch Report on the Tactile Internet, which emphasised that scaling up research in the area of wired and wireless access networks would be essential, ushering in new ideas and concepts to boost access networks' redundancy and diversity to meet the stringent latency as well as carrier-grade reliability requirements of Tactile Internet applications [22]. The Tactile Internet provides a medium for remote physical interaction in real-time, which requires the exchange of closed-loop information between virtual or real objects (i.e., humans, machines, and processes). This mandatory end-to-end design approach is fully reflected in the key principles of the reference architecture within the emerging IEEE P1918.1 standards working group (formed in March 2016), which aims to define a framework for the Tactile Internet [23]. Among others, the key principles envision to (i) develop a generic Tactile Internet reference architecture, (ii) support local area as well as wide area connectivity through wireless (e.g., cellular, WiFi) or hybrid wireless and wired networking, and (iii) leverage computing resources from cloud variants at the edge of the network. The working group defines the Tactile Internet as a 'network or network of networks for remotely accessing, perceiving, manipulating or controlling real or virtual objects or processes in perceived real-time by humans or machines'. Some of the key use cases considered in IEEE P1918.1 include teleoperation, haptic communications, immersive virtual reality, and automotive control.

To give it a more 5G-centric flavour, the Tactile Internet has been more recently also referred to as the 5G-enabled Tactile Internet [24,25]. Recall that unlike the previous four cellular generations, future 5G networks will lead to an increasing integration of cellular and WiFi technologies and standards [26]. Furthermore, the importance of the so-called backhaul bottleneck needs to be recognised as well, calling for an end-to-end design approach leveraging both wireless front-end and wired backhaul technologies. Or, as eloquently put by Andrews, the lead author of [26], 'placing base stations all over the place is great for providing the mobile stations high-speed access, but does this not just pass the buck to the base stations, which must now somehow get this data to and from the wired core network?' [27].

Clearly, the Tactile Internet opens up a plethora of exciting research directions towards adding a new dimension to the human-to-machine interaction via the Internet. According to the aforementioned ITU-T Technology Watch Report, the Tactile Internet is supposed to be the next leap in the evolution of today's IoT, although there is a significant overlap among 5G, IoT, and the Tactile Internet. For illustration, Figure 1.3 provides a view of the aforementioned commonalities and differences through the three lenses of IoT, 5G, and the Tactile Internet. The major differences may be best expressed in terms of underlying communications paradigms and enabling devices. IoT relies on M2M communications with a focus on smart devices (e.g., sensors and actuators). In coexistence with emerging MTC, 5G will maintain its traditional human-to-human (H2H) communications paradigm for conventional triple-play services (voice, video, and data) with a growing focus on the integration with other wireless technologies (most notably WiFi) and decentralisation. Conversely, the Tactile Internet will be

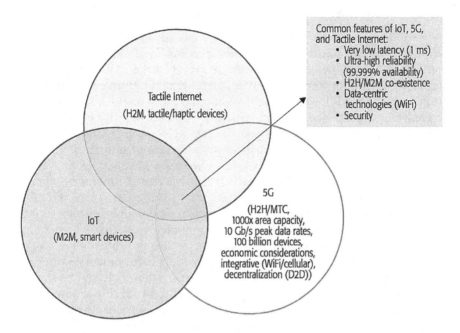

Figure 1.3 The three lenses of the Internet of Things (IoT), 5G, and the Tactile Internet: Commonalities and differences.

centered on human-to-machine (H2M) communications leveraging tactile and haptic devices. More importantly, despite their differences, IoT, 5G, and the Tactile Internet seem to converge toward a common set of important design goals:

- Very low latency on the order of 1 ms
- Ultra-high reliability with an almost guaranteed availability of 99.999%
- H2H and M2M coexistence
- Integration of data-centric technologies with a particular focus on WiFi
- Security

Importantly, the Tactile Internet involves the inherent human-in-the-loop (HITL) nature of H2M interaction, as opposed to the emerging IoT without any human involvement in its underlying M2M communications. Although M2M communications is useful for the automation of industrial and other machine-centric processes, the Tactile Internet will be centered on human-to-machine/robot (H2M/R) communications and will thus allow for a human-centric design approach towards creating novel immersive experiences and extending the capabilities of the human through the Internet (i.e., augmentation rather than automation of the human) [28].

1.4.2 Human-agent-robot teamwork

A promising approach towards achieving advanced human-machine coordination by means of a superior process for fluidly orchestrating human and machine coactivity may be found in the still young field of human-agent-robot teamwork (HART) research [29]. Unlike early automation research, HART goes beyond the singular focus on full autonomy (i.e., complete independence and self-sufficiency) and cooperative/collaborative autonomy among autonomous systems themselves, which aim at excluding humans as potential teammates for the design of human-out-of-the-loop solutions. In HART, the dynamic allocation of functions and tasks between humans and machines, which may vary over time or be unpredictable in different situations, plays a central role. In particular, with the rise of increasingly smarter machines, the historical humans-are-better-at/machines-are-better-at (HABA/MABA) approach to decide which tasks are best performed by people and which by machines rather than working in concert has become obsolete. To provide a better understanding of the potential and limitations of current smart machines, T. H. Davenport and J. Kirby classified the capabilities of intelligent machines along two dimensions, namely their ability to act and their ability to learn, in their book, *Only Humans Need Apply: Winners and Losers in the Age of Smart Machines*. The ability to act involves four task levels, ranging from the most basic tasks (e.g., analysing numbers) to performing digital tasks (done by agents) or even physical tasks (done by robots). On the other hand, the ability to learn escalates through four levels, spanning from human-support machines with no inherent intelligence to machines with context awareness, learning, or even self-aware intelligence.

According to [29], among other HART research challenges, the development of capabilities that enable autonomous systems not merely to do things for people but also to work together with people and other systems represents an important open issue to treat humans as 'members' of a team of intelligent actors rather than keep viewing them as a conventional 'users'. In the following, we introduce and extend the concept of fibre-wireless (FiWi) enhanced LTE-advanced (LTE-A) heterogeneous networks (HetNets) to enable both local and non-local teleoperation by exploiting AI-enhanced MEC capabilities to achieve both low round-trip latency and low jitter.

1.4.3 Low-latency FiWi-enhanced LTE-A HetNets with AI-enhanced MEC

1.4.3.1 Low-latency FiWi-enhanced LTE-A HetNets

FiWi access networks, also referred to as 'wireless-optical broadband access networks' (WOBANs), combine the reliability, robustness, and high capacity of optical fibre networks and the flexibility, ubiquity, and cost savings of wireless networks [30]. To deliver peak data rates up to 200 Mbps per user and realise the vision of complete fixed-mobile convergence, it is crucial to replace today's legacy wireline and microwave backhaul technologies with integrated FiWi broadband access networks.

In [31], we investigated the performance gains obtained from unifying coverage-centric 4G mobile networks and capacity-centric FiWi broadband access networks based on data-centric Ethernet technologies with resulting fibre backhaul sharing and WiFi offloading capabilities in response to the unprecedented growth of mobile data traffic. We evaluated the maximum aggregate throughput, offloading efficiency, and in particular, the delay performance of FiWi-enhanced LTE-A HetNets, including the beneficial impact of various localised fibre-lean backhaul redundancy and wireless protection techniques, by means of probabilistic analysis and verifying simulation. In our study, we paid close attention to fibre backhaul reliability issues stemming from fibre faults of an Ethernet passive optical network (EPON) and WiFi offloading limitations resulting from WiFi mesh node failures as well as temporal and spatial WiFi coverage constraints.

For illustration, Figure 1.4 depicts the average end-to-end delay performance of FiWi-enhanced LTE-A Het- Nets versus aggregate throughput for different WiFi offloading ratio (WOR) values, whereby $0 \leq WOR \leq 1$

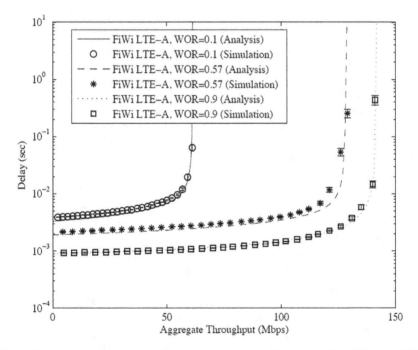

Figure 1.4 Average end-to-end delay versus aggregate throughput for different WiFi offloading ratio (WOR).

denotes the percentage of mobile user traffic offloaded onto WiFi. The presented analytical and verifying simulation results were obtained by assuming a realistic LTE-A and FiWi network configuration under uniform traffic loads and applying minimum (optical and wireless) hop routing. For further details the interested reader is referred to [31]. For now, let us assume that the reliability of the EPON is ideal (i.e., no fibre backhaul faults occur). However, unlike EPON, the WiFi mesh network may suffer from wireless service outage with probability 10^{-6}. We observe from Figure 1.4 that for increasing WOR the throughput-delay performance of FiWi-enhanced LTE-A HetNets is improved significantly. More precisely, by changing WOR from 0.1 to 0.57 the maximum achievable aggregate throughput increases from about 61 Mbps to roughly 126 Mbps (i.e., the maximum achievable aggregate throughput has more than doubled). More importantly, further increasing WOR to 0.9 does not result in an additional significant increase of the maximum achievable aggregate throughput, but it is instrumental in decreasing the average end-to-end delay and keeping it at a very low level of 10^{-3} second (1 ms) for a wide range of traffic loads. Thus, this result shows that WiFi offloading the majority of data traffic from 4G mobile networks is a promising approach to obtain a very low latency on the order of 1 ms.

1.4.3.2 AI-enhanced MEC: Pushing AI to the edge

Figure 1.5 depicts the generic network architecture of FiWi-enhanced LTE-A HetNets with AI-enhanced MEC server. The fibre backhaul consists of a time or wavelength division multiplexing (TDM/WDM) IEEE 802.3ah/av 1/10 Gb/s EPON with a typical fibre range of 20 km between the central optical line terminal (OLT) and remote optical network units (ONUs). The EPON may comprise multiple stages, each stage separated by a wavelength-broadcasting splitter and combiner or a wavelength multiplexer and demultiplexer. There are three different subsets of ONUs. An ONU may either serve fixed (wired) subscribers. Alternatively, it may connect to a cellular network base station (BS) or an IEEE 802.11n/ac/s WLAN mesh portal point (MPP), giving rise to a collocated ONU-BS or ONU-MPP, respectively. Depending on the trajectory, a mobile user (MU) may communicate through the cellular network or WLAN mesh front end, which consists of ONU-MPPs, intermediate mesh points (MPs), and mesh access points (MAPs) [32].

Human operators (HOs) and teleoperator robots (TORs) are assumed to communicate only via WLAN, as opposed to MUs using their dual-mode 4G/WiFi smartphones. Teleoperation is done either locally or non-locally, depending on the proximity of the involved HO and TOR, as illustrated in Figure 1.5. In local teleoperation, the HO and corresponding TOR are associated with the same MAP and exchange their command and feedback samples through this MAP without traversing the fibre backhaul. Conversely, if HO and TOR are associated with different MAPs, non-local

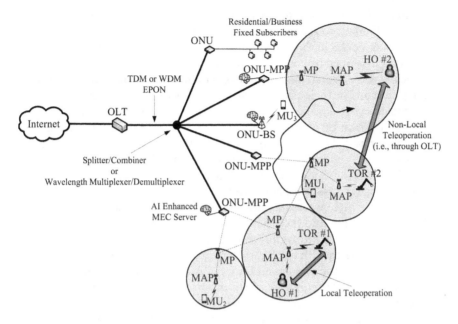

Figure 1.5 Architecture of FiWi-based Tactile Internet with artificial intelligence (AI)-enhanced MEC for local and non-local teleoperation.

teleoperation is generally done by communicating via the backhaul EPON and central OLT. Despite recent interest in exploiting machine learning for optical communications and networking, edge intelligence for enabling an immersive and transparent teleoperation experiences for HOs has not been explored yet. In [32], we applied machine learning at the edge of our considered communication network for realising immersive and frictionless Tactile Internet experiences. To realise edge intelligence, selected ONU-BSs/MPPs are equipped with AI-enhanced MEC servers. These servers rely on the computational capabilities of cloudlets collocated at the optical-wireless interface (see Figure 1.5) to forecast delayed haptic samples in the feedback path. As a consequence, the HO is enabled to perceive the remote task environment in real-time at a 1-ms granularity, thus resulting in a tighter togetherness, improved safety control, and increased reliability of the teleoperation systems.

1.5 DECENTRALISING THE TACTILE INTERNET

Recall from Section 1.1 that the Tactile Internet will involve the inherent HITL nature of H2M interaction, as opposed to the emerging IoT. Thus, it allows for a human-centric design approach towards extending the capabilities of the human through the Internet for the augmentation rather than automation of the human. Recently, in [28], we put forward the idea that the Tactile Internet may be the harbinger of human augmentation and human-machine symbiosis envisioned by contemporary and early-day Internet pioneers. More specifically, we elaborated on the role of AI-enhanced agents may play in supporting humans in their task coordination between humans and machines. Toward achieving advanced human-machine coordination, we developed a distributed allocation algorithm of computational and physical tasks for fluidly orchestrating HART coactivities (e.g., the shared use of user- and/or network-owned robots). In our design approach, all HART members established through communication a collective self-awareness with the objective of minimising the task completion time.

In the following, we search for synergies between the aforementioned HART membership and the complementary strengths of the DAO, AI, and robots (see Figure 1.2) to facilitate local human-machine coactivity clusters by decentralising the Tactile Internet. Towards this end, it is important to better understand the merits and limits of AI. Recently, Stanford University launched its *One Hundred Year Study on Artificial Intelligence (AI100)*. In the inaugural report 'Artificial Intelligence and Life in 2030,' the authors defined AI as a set of computational technologies that are inspired by the ways people use their nervous systems and bodies to sense, learn, reason, and take action. They also point out that AI will likely replace tasks rather than jobs in the near term and highlight the importance of crowdsourcing of human expertise to solve problems that computers alone cannot solve well.

1.5.1 Decentralised edge intelligence

First, let us explore the potential of leveraging mobile end-user equipment by partially or fully decentralising MEC. We introduce the use of AI-enhanced MEC servers at the optical-wireless interface of converged fibre-wireless mobile networks for computation offloading. Assuming the same default network parameter setting and simulation setup as in [28], we consider $1 \leq N_{Edge} \leq 4$ AI-enhanced MEC servers, each associated with 8 end users, whereof $1 \leq N_{PD} \leq 8$ partially decentralised end users can flexibly control the amount of offloaded tasks by varying their computation offloading probability. The remaining $8 - N_{PD}$ are fully centralised end users that rely on edge computing only (i.e., their computation offloading probability equals 1). Note that for $N_{Edge} = 4$, all end users may offload their computation tasks onto an edge node. Conversely, for $N_{Edge} < 4$, one or more edge nodes are unavailable for computation offloading and their associated end users fall back on their local computation resources (i.e., fully decentralised). Figure 1.6 shows the average task completion time versus computation offloading probability of the partially decentralised end users for different N_{Edge} and N_{PD}. We observe from Figure 1.6 that for a given N_{Edge}, increasing N_{PD} (i.e., higher level of decentralisation) is effective

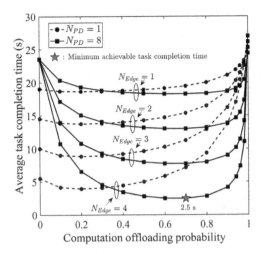

Figure 1.6 Average task completion time (in seconds) versus computation offloading probability for different numbers of partially decentralised end users (N_{PD}) and artificial intelligence (AI)-enhanced MEC servers (N_{Edge}).

in reducing the average task completion time. Specifically, for $N_{Edge} = 4$, a high decentralisation level ($N_{PD} = 8$) allows end users to experience a reduction of the average task completion time of up to 89.5% by optimally adjusting their computation offloading probability to 0.7.

As interconnected computing power has spread around the world and useful platforms have been built on top of it, the crowd has become a demonstrably viable and valuable resource. According to [6], there are many ways for companies that are squarely at the core of modern capitalism to tap into the expertise of uncredentialed and conventionally inexperienced members of the technology-enabled crowd such as the DAO. In [28], we developed a self-aware allocation algorithm of physical tasks for HART-centric task coordination based on the shared use of user- and network-owned robots. Recall from Section 1.4.3.2 that by using our AI-enhanced MEC servers as autonomous agents, we showed that delayed force feedback samples coming from TORs may be locally generated and delivered to HOs in close proximity. Note, however, that the performance of the sample forecasting-based teleoperation system heavily relies on the accuracy of the forecast algorithm.

1.5.2 Crowdsourcing: Expanding the HO workforce

Towards realising DAO in a decentralised Tactile Internet, Ethereum may be used to establish HO-TOR sessions for remote physical task execution, whereby smart contracts help establish and maintain trusted HART membership and allow each HART member to have global knowledge about all participating HOs, TORs, and MEC servers that act as autonomous agents

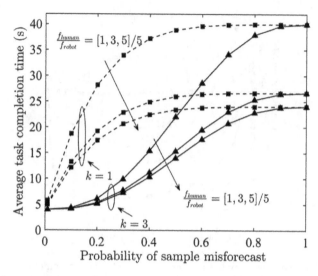

Figure 1.7 Average task completion time (in seconds) versus the probability of sample misforecast for different number of misforecasts (k) and ratio of human and robot operational capabilities (f_{human}/f_{robot}).

following the widely used gateway-oriented approach (see Section 1.3.1). An HO remotely executes a given physical task until k out of the recent n haptic feedback samples are misforecast. At this point, the HO immediately stops the teleoperation and informs the agent. The agent assigns the interrupted task to an available HO in vicinity, who then traverses to the task point and finalises the physical task. For $n = 5$, Figure 1.7 depicts the average task completion time versus probability of sample misforecast for different k and f_{human}/f_{robot}, where f_{human} and f_{robot} denote the capability (in terms of number of operations per second) of a human and robot for performing physical tasks, respectively. We observe that the task completion time increases as the probability of sample misforecast grows from 0 to 0.25 for $k = 1$, whereas it remains almost unchanged for $k = 3$. Further, note that as humans become more capable (i.e., increasing f_{human}/f_{robot}), the resultant task completion time decreases until it hits a plateau, as traversing incurs additional delays.

1.6 BLOCKCHAIN, AI, AND HUMAN INTELLIGENCE: THE PATH FORWARD

Today, AI, machine learning, robots, and IoT are converging to create a new wave of change as they begin to take advantage of cryptocurrencies, initial coin offerings (ICOs), virtual assets, and tokenisation of everything.

The outcome of this convergence of innovations and human cause fears about job loss, robot overlords, and the future in general [33].

In [28], we touched on the importance of shifting the research focus from AI to intelligence amplification (IA) by using information technology to enhance human decisions. Note, however, that IA becomes difficult in dynamic task environments of increased uncertainty and real-world situations of great complexity. IA, also known as 'cognitive augmentation' or 'machine-augmented intelligence', will be instrumental in enhancing the creativity, understanding, efficiency, and intelligence of humans.

1.6.1 Cognitive-assistance-based intelligence amplification

Edge computing is a widely studied approach to increase the usefulness of crowdsourcing, which may be used to guide humans step by step through the physical task execution process by providing them with cognitive assistance. Technically, this could be easily realised by equipping an unskilled DAO member with an augmented reality (AR) headset (e.g., HoloLens 2 with WiFi connectivity) that receives work-order information in real-time from its nearest AI-enhanced MEC server.

1.6.2 HITL hybrid-augmented intelligence

Many problems that humans face tend to be of high uncertainty, complexity, and open-ended. To solve such problems, human interaction and participation may be introduced. As a result, this gives rise to the concept of HITL hybrid-augmented intelligence for advanced human-machine collaboration [34]. HITL hybrid-augmented intelligence is defined as an intelligent model that requires human interaction and allows for addressing problems and requirements that may not be easily trained or classified by machine learning, In general, machine learning is inferior to the human brain in understanding unstructured real-world environments and processing incomplete information and complex spatio-temporal correlation tasks. Hence, machines cannot carry out all the tasks in human society on their own. Instead, AI and human intelligence are better viewed as highly complementary.

According to [34], the Internet provides an immense innovation space for HITL hybrid-augmented intelligence. Specifically, cloud robotics and AR are among the fastest-growing commercial applications for enhancing the intelligence of an individual in multi-robot collaborative systems. One of the main research topics of HITL hybrid-augmented intelligence is the development of methods that allow machines to learn from not only massive training samples but also human knowledge to accomplish highly intelligent tasks via shared intelligence among different robots and humans.

1.6.3 The rise of the decentralised self-organising cooperative

A very interesting example to catalyse human and machine intelligence towards a new form of self-organising artificial general intelligence (AGI) across the Internet is the so-called SingularityNET.[2] One can think of SingularityNet as a decentralised self-organising cooperative (DSOC), a concept similar to the aforementioned DAO. DSOC is essentially a distributed computing architecture for making new kinds of smart contracts. Entities executing these smart contracts are referred to as 'agents', which can run in the cloud, on phones, robots, or other embedded devices. Services are offered to any customer via APIs enabled by smart contracts and may require a combination of actions by multiple agents using their collective intelligence. In general, there may be multiple agents that can fulfill a given task request in different ways and to different degrees. Each task request to the network requires a unique combination of agents, thus forming a so-called' offer network of mutual dependency', where agents make offers to each other to exchange services via offer-request pairs. Whenever someone wants an agent to perform services, a smart contract is signed for this specific task. Towards this end, DSOC aims at leveraging contributions from the broadest possible variety of agents by means of superior discovery mechanisms for finding useful agents and nudging them to become contributors.

1.6.4 Nudging towards human augmentation

Extending on the DSOC concept, we advocate the use of the nudge theory for enhancing the human capabilities of unskilled crowd members of the DAO. According to Richard H. Thaler, the 2017 Nobel Laureate in Economics, the nudge theory claims that positive reinforcement and indirect suggestions can influence the behaviour of groups and individuals. A suitable nudging mechanism aims at completing interrupted local physical tasks by learning from a remote skilled DAO member, who is able to transfer her knowledge as data via a secure blockchain transaction embedded in a smart contract with an appropriate reward for each subtask. This smart contract will initially enhance the local HO's capability, thereby allowing to successfully accomplish a given task via shared intelligence among failing robots and skilled humans.

1.7 OPEN CHALLENGES AND FUTURE WORK

The Internet has been constantly evolving from the mobile Internet to the emerging IoT and future Tactile Internet. Similarly, the capabilities of future 5G networks will extend far beyond those of previous generations

---

[2] SingularityNET Whitepaper, 'SingularityNET: A decentralized, open market and internetwork for AIs,' November 2017. https://singularitynet.io.

of mobile communication. By boosting access networks' redundancy and diversity as envisioned in [22], FiWi-enhanced 4G LTE-A HetNets with AI-embedded MEC hold great promise to meet the stringent latency and reliability requirements of immersive Tactile Internet applications. In [35], we recently outlined some research ideas that help tap into the full potential of the Tactile Internet. However, other concepts such as the discussed spreading ownership, DAO, and HART membership will be instrumental in ushering in new ideas and concepts to facilitate local human-machine coactivity clusters by completely decentralising edge computing via emerging Ethereum blockchain technologies to realise future techno-social systems such as the Tactile Internet, which by design still requires heavy involvement from humans at the network edge instead of automating them away. More work lies ahead to integrate Ethernet-based FiWi-enhanced mobile networks with Ethereum blockchain technologies. Although blockchain technology is promising technology to enhance today's IoT, there are still many research issues to be addressed before the integration of blockchain with IoT, especially in future mobile networks (i.e., 6G and beyond) that play a primordial role in constructing the underlying infrastructure for blockchains.

1.8 CONCLUSIONS

In this chapter, we showed that many of the emerging B-IoT studies use Ethereum as the blockchain of choice and apply a gateway-oriented design approach to offload computationally intensive tasks from resource-constrained end devices onto an intermediate gateway, thus enabling them to access the Ethereum blockchain network. Building on our recent Tactile Internet work on orchestrating hybrid HART coactivities, we showed that higher levels of decentralised AI-enhanced MEC are effective in reducing the average completion time of computational tasks. Further, for remote execution of physical tasks in a decentralised Tactile Internet, we explored how Ethereum's DAO and smart contracts may be used to establish trusted HART membership and how human crowdsourcing helps decrease physical task completion time in the event of unreliable forecasting of haptic feedback samples from teleoperated robots. We outlined future research avenues on technological convergence to successfully accomplish hybrid machine-human tasks by tapping into the shared intelligence of the crowd.

REFERENCES

1. M. Maier, M. Chowdhury, B. P. Rimal and D. P. Van, 'The Tactile Internet: Vision, Recent Progress, and Open Challenges,' *IEEE Communications Magazine*, vol. 54, no. 5, pp. 138–145, 2016.

2. D. Tapscott and A. Tapscott, *Blockchain Revolution: How the Technology Behind Bitcoin Is Changing Money, Business, and the World*, Portfolio, Toronto, Canada, 2016.
3. W. Saad, M. Bennis and M. Chen, 'A Vision of 6G Wireless Systems: Applications, Trends, Technologies, and Open Research Problems,' *IEEE Network*, vol. 34, no. 3, pp. 134–142, May/June 2020.
4. T. Salman, M. Zolanvari, A. Erbad, R. Jain and M. Samaka, 'Security Services Using Blockchains: A State of the Art Survey,' *IEEE Communications Surveys & Tutorials*, vol. 21, no. 1, pp. 858–880, 2019.
5. V. Buterin, 'A Next-Generation Smart Contract and Decentralized Application Platform,' *Ethereum White Paper*, www.ethereum.org.
6. A. McAfee and E. Brynjolfsson, *Machine, Platform, Crowd: Harnessing Our Digital Future*, W. W. Norton, New York, 2017.
7. F. Tschorsch and B. Scheuermann, 'Bitcoin and Beyond: A Technical Survey on Decentralized Digital Currencies,' *IEEE Communications Surveys & Tutorials*, vol. 18, no. 3, pp. 2084–2123, 2016.
8. R. Beck, 'Beyond Bitcoin: The Rise of Blockchain World,' *IEEE Computer*, vol. 51, no. 2, pp. 54–58, 2018.
9. O. Novo, 'Blockchain Meets IoT: An Architecture for Scalable Access Management in IoT,' *IEEE Internet of Things Journal*, vol. 5, no. 2, pp. 1184–1195, 2018.
10. Y. N. Aung and T. Tantidham, 'Review of Ethereum: Smart Home Case Study,' *Proceedings of the 2nd International Conference on Information Technology (INCIT)*, Nakhon Pathom, Thailand, 2017, pp. 1–4.
11. C.-F. Liao, S.-W. Bao, C.-J. Cheng and K. Chen, 'On Design Issues and Architectural Styles for Blockchain-Driven IoT Services,' *Proceedings of the IEEE International Conference on Consumer Electronics-Taiwan (ICCE-TW)*, Taipei, 2017, pp. 351–352.
12. P. Urien, 'Towards Secure Elements for Trusted Transactions in Blockchain and Blockchain IoT (BIoT) Platforms,' *Proceedings of the Fourth International Conference on Mobile and Secure Services (MobiSecServ)*, Miami Beach, FL, 2018, pp. 1–5.
13. G. C. Polyzos and N. Fotiou, 'Blockchain-Assisted Information Distribution for the Internet of Things,' *Proceedings of the IEEE International Conference on Information Reuse and Integration (IRI)*, San Diego, CA, 2017, pp. 75–78.
14. Y. Gupta, R. Shorey, D. Kulkarni and J. Tew, 'The Applicability of Blockchain in the Internet of Things,' *Proceedings of the 10th International Conference on Communication Systems & Networks (COMSNETS)*, Bengaluru, India, 2018, pp. 561–564.
15. K. R. Özyilmaz and A. Yurdakul, 'Work-in-Progress: Integrating Low-Power IoT Devices to a Blockchain-Based Infrastructure,' *Proceedings of the International Conference on Embedded Software (EMSOFT)*, Seoul, South Korea, 2017, pp. 1–2.
16. X. Duan, Z. Yan, G. Geng and B. Yan, 'DNSLedger: Decentralized and Distributed Name Resolution for Ubiquitous IoT,' *Proceedings of the IEEE International Conference on Consumer Electronics (ICCE)*, Las Vegas, NV, 2018, pp. 1–3.

17. R. Yang, F. R. Yu, P. Si, Z. Yang and Y. Zhang, 'Integrated Blockchain and Edge Computing Systems: A Survey, Some Research Issues and Challenges,' *IEEE Communications Surveys & Tutorials*, vol. 21, no. 2, pp. 1508–1532, 2019.

18. J. Pan, J. Wang, A. Hester, I. Alqerm, Y. Liu and Y. Zhao, 'EdgeChain: An Edge-IoT Framework and Prototype Based on Blockchain and Smart Contracts,' *IEEE Internet of Things Journal*, vol. 6, no. 3, pp. 4719–4732, 2019.

19. X. Xu, X. Zhang, H. Gao, Y. Xue, L. Qi and W. Dou, 'BeCome: Blockchain-Enabled Computation Offloading for IoT in Mobile Edge Computing,' *IEEE Transactions on Industrial Informatics*, vol. 16, no. 6, pp. 4187–4195, 2020, doi:10.1109/TII.2019.2936869.

20. M. Zhaofeng, W. Xiaochang, D. K. Jain, H. Khan, G. Hongmin and W. Zhen, 'A Blockchain-Based Trusted Data Management Scheme in Edge Computing,' *IEEE Transactions on Industrial Informatics*, vol. 16, no. 3, pp. 2013–2021, 2020, doi:10.1109/TII.2019.2933482.

21. G. P. Fettweis, 'The Tactile Internet: Applications and Challenges,' *IEEE Vehicular Technology Magazine*, vol. 9, no. 1, pp. 64–70, 2014.

22. 'ITU-T, 'The Tactile Internet,' International Telecommunication Union (ITU), Technology Watch Report, August 2014. [Online]. Available: https://www.itu.int/dms pub/itu-t/oth/23/01/T23010000230001PDFE.pdf.

23. A. Aijaz, Z. Dawy, N. Pappas, M. Simsek, S. Oteafy and O. Holland, 'Toward a Tactile Internet Reference Architecture: Vision and Progress of the IEEE P1918.1 Standard,' *arXiv:1807.11915*, 2018.

24. M. Simsek, A. Aijaz, M. Dohler, J. Sachs and G. Fettweis, '5G-enabled Tactile Internet,' *IEEE Journal of Selected Areas in Communications*, vol. 34, no. 3, pp. 460–473, 2016.

25. A. Aijaz, M. Dohler, A. H. Aghvami, V. Friderikos and M. Frodigh, 'Realizing the Tactile Internet: Haptic Communications Over Next Generation 5G Cellular Networks,' *IEEE Wireless Communications*, vol. 24, no. 2, pp. 82–89, 2017.

26. J. G. Andrews, S. Buzzi, W. Choi, S. V. Hanley, A. Lozano, A. C. K. Soong and J. C. Zhang, 'What Will 5G Be?' *IEEE Journal of Selected Areas in Communications*, vol. 32, no. 6, pp. 1065–1082, 2014.

27. J. G. Andrews, 'Seven Ways that HetNets Are a Cellular Paradigm Shift,' *IEEE Communications Magazine*, vol. 51, no. 3, pp. 136–144, 2013.

28. M. Maier, A. Ebrahimzadeh and M. Chowdhury, 'The Tactile Internet: Automation or Augmentation of the Human?' *IEEE Access*, vol. 6, pp. 41607–41618, 2018.

29. J. M. Bradshaw, V. Dignum, C. Jonker and M. Sierhuis, 'Human-Agent-Robot Teamwork,' *IEEE Intelligent Systems*, vol. 27, no. 2, pp. 8–13, 2012.

30. F. Aurzada, M. Lévesque, M. Maier and M. Reisslein, 'FiWi Access Networks Based on Next-Generation PON and Gigabit-Class WLAN Technologies: A Capacity and Delay Analysis,' *IEEE/ACM Transactions on Networking*, vol. 22, no. 4, pp. 1176–1189, 2014.

31. H. Beyranvand, M. Lévesque, M. Maier and J. A. Salehi, 'FiWi Enhanced LTE-A HetNets with Unreliable Fiber Backhaul Sharing and WiFi Offloading,' *Proceedings of the IEEE INFOCOM*, Hong Kong, 2015, pp. 1275–1283.

32. M. Maier and A. Ebrahimzadeh, 'Towards Immersive Tactile Internet Experiences: Low-Latency FiWi Enhanced Mobile Networks with Edge Intelligence [Invited],' *IEEE/OSA Journal of Optical Communications and Networking, Special Issue on Latency in Edge Optical Networks*, vol. 11, no. 4, pp. B10–B25, 2019.

33. A. Pentland, J. Werner and C. Bishop, 'Blockchain+AI+Human: Whitepaper and Invitation,' MIT Press, 2018. [Online]. Available: https://connection.mit.edu/sites/default/files/publicationpdfs/blockchain%2BAI%2BHumans.pdf.

34. N. Zheng, Z. Liu, P. Ren, Y. Ma, S. Chen, S. Yu, J. Xue, B. Chen and F. Wang, 'Hybrid-Augmented Intelligence: Collaboration and Cognition,' *Frontiers of Information Technology & Electronic Engineering*, vol. 18, no. 2, pp. 153–179, 2017.

35. M. Maier, 'The Tactile Internet: Where Do We Go from Here? (Invited Paper),' *IEEE/OSA/SPIE Asia Communications and Photonics (ACP) Conference*, Hangzhou, China, 2018.

Chapter 2

Blockchain concepts, architectures, and Smart city applications in fog and edge computing environments

Daniel Minoli and Benedict Occhiogrosso

CONTENTS

2.1 INTRODUCTION AND OVERVIEW

There are three evolving networking technologies that will invariably play pivotal worldwide roles in the decade of the 2020s: (i) Internet of Things (IoT) capabilities for a large array of endpoints and objects; (ii) high-capacity 5G cellular connectivity, especially in urban and suburban environments; and (iii) enhanced cybersecurity mechanisms in general and blockchain-protected transactions in particular. Many evolving environments will subsume all three technologies. Another trend is the emergence of Smart cities. This chapter discusses the synergistic value of this combination of new tools and services for an ever-enlarging set of business and personal applications, particularly in the Smart city environment. New York City is typically considered to be a top-tier Smart city, and as professionals working in this geographic area, the authors offer some pragmatic views of these evolving disciplines based on several projects they are involved with. The focus of this chapter is how the IoT and 5G specifically can interplay and interact and how IoT applications can use blockchain technologies.

Following a motivation section (Section 2.2), this chapter provides a brief assessment of IoT architectures with emphasis on the edge structures (Section 2.3). This is followed by a short discussion of applicable wireless communication technologies, including emerging 5G cellular services (Section 2.4). Many IoT applications will initially focus on Smart city environments; therefore, a description of Smart city concepts and common Smart city applications is provided (Section 2.5). A short summary description of applicable blockchain concepts follows (Section 2.6). Next, a more detailed survey of Smart city IoT applications that are or will benefit from blockchains, with emphasis on the edge functionality, is provided, highlighting the following applications realms: e-health, logistics and supply chain, surveillance, and 'ring of steel', municipal smart services, insurance, and banking (Section 2.7). An assessment of future research areas is also provided (Section 2.8).

2.2 BACKGROUND AND MOTIVATION

2.2.1 IoT ecosystem

IoT applications span many realms and range from supporting mission-critical environments (e.g., intelligent transportation systems [ITSs], smart grid, industrial process control, video surveillance, and e-health), to facilitating business-oriented transactions (e.g., contracts, insurance, banking, and various logistics). To support these applications, various types of sensors are dispersed in the environment; these sensors can be stationary or, more typically, mobile. Often, but not always, sensors rely on (limited) on-board power, in the form of small batteries intended to last up to 10 years. Very often, but not always, sensors use wireless technologies to forward their data to an aggregation point or, possibly, to a final destination point, such as an analytics engine; typically, low power wide area network (LPWAN) technologies are used. Edge aggregation in the form of a routing point or switching node possibly also performing computing and processing – for example, data correlation, reduction, or accumulation – is often used; this hierarchical model not only results in fewer nodes accessing the cloud-resident analytics directly, thus reducing network traffic, but also supports lower-power transmission (extending, battery life), by requiring just enough transmission power to reach a neighbourhood (or neighbouring) node [1,2]. Considering the broad applicability of the technology it should not be a surprise that there is an extensive body of literature of the topic of 'IoT' AND 'Blockchain' AND 'Application'; for example, a Google Scholar search on that term combination identifies more than 8,000 technical papers (more than 1,700 just in 1H2019).

2.2.2 Edge computing in the IoT ecosystem

For the purpose of this paper, fog computing and fog networking entail content-oriented computing and communication capabilities at a local hub device, including the case where these capabilities are embedded directly into the smart object itself. Edge networking and computing entail the large-scale networking aggregation and possibly also some additional content-oriented computing of a set of local or dispersed devices that support fog computing. A fog network is also referred to as 'a fog'. Effectively, fog computing entails the 'insertion' of a new layer between cloud software computing resources and IoT devices [3]. Fog computing is a paradigm that uses a group of fog nodes that are located at the edge, in proximity of the end nodes to offload some of the functionality originally provided in data centres resident in the cloud. In some cases, these edge and gateway nodes have relatively low-computing power available locally; in other cases, the edge devices are able to support substantial amount of computation, storage, and local communication. One can refer to the fog network as the set

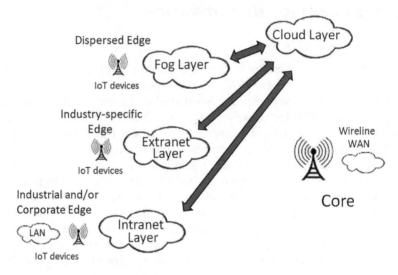

Figure 2.1 Various edge and fog arrangements.

of device-facing communication links that allow remote dispersed IoT nodes to communicate with the fog nodes [4–13]. In some environments or implementations, the fog functionality and the edge functionality are co-located or are logically merged; however, in other environments, they are plainly distinct.

Figure 2.1 depicts schematically some examples of edge environments and fogs. Some IoT applications require fully distributed IoT devices that are broadly dispersed over the open environment; these systems often entail an explicit fog layer where concentration, protocol conversion, processing, transient storage, and other functions may take place, prior to having the data transiting a wide area network (WAN) to reach the service-rich cloud. Some IoT applications entail an industry-specific set of IoT devices, for example in the case of a smart grid or a transportation rail system; these systems may be aggregated at the edge but typically use an industry-specific traditional network (shown as 'extranet' in the figure). Other applications are more geographically or administratively localised, such as an industrial IoT environment or a corporate IoT environment – for example a building management system (BMS); these systems may employ a localised wired or wireless LAN for aggregation and access to the WAN and cloud and may use a fog layer mechanism per se.

Figure 2.2 depicts an illustrative example of the fog and edge-oriented physical topology of a Smart city IoT ecosystem, which, in fact, often entails a fully distributed set of IoT devices. In this illustrative example, nine fogs or edge networks are shown. For several fogs (e.g., Fog 1, Fog 2, Fog 3, Fog 4, etc.), the communication is directly with a nearby (cellular) tower (e.g., for LTE-M or NB-IoT services); the fog computing function is deeper in the access network, for example it could be at the mobile central office (MCO). In Fog 2 the nearby tower is connected to the MCO via a fiberoptic link; in Fog 3 the

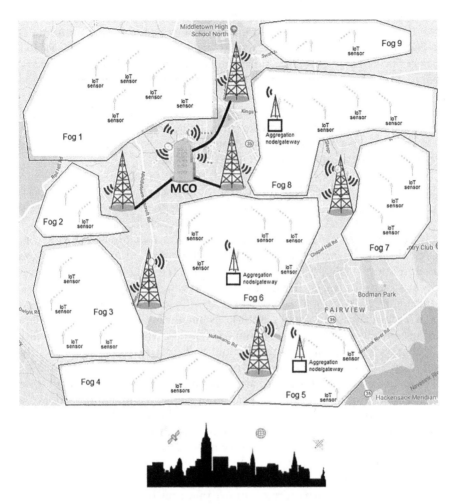

Figure 2.2 Illustrative example of the fog- and edge-oriented physical topology of a Smart city Internet of Things (IoT).

nearby tower is connected to the MCO via a microwave link. Several fogs (e.g., Fog 5, Fog 6, Fog 8) use a strictly local aggregation or gateway function that may supply network processing (e.g., multiplexing or firewalling) or a data-processing function (e.g., data summarisation, averaging, temporary storage) prior to reaching the inner layers of the core or cloud network.

2.2.3 IoT security concepts

There are some basic motivating factors for considering blockchain mechanisms (BCM) in the IoT and Smart city environment: Given the torrent of emerging IoT applications there is now clear awareness that cybersecurity is of critical importance in the IoT. Just as is the case in information technology (IT) domains, there is a need in IoT to detect and remediate routine as well as advanced persistent threats (APTs) that may be placed against IoT devices or systems. Given their distributed, dynamic, and highly connected paradigm, IoT systems present an enlarged cyberattack surface. Security and privacy threats in IoT-enabled applications include, among others, forgery, tampering, data privacy, Sybil attacks, eavesdropping, jamming, denial of service, man-in-the-middle attacks, impersonation, identity privacy exposure, usage privacy exposure, location privacy exposure, and corrupted operating systems and application software uploads.

There is, therefore, a need for thorough support of security in the IoT, particularly for mission-critical applications, also including the affiliated business analytics that are downstream from the sensors themselves. The field of IoT Security (IoTSec) is now beginning to coalesce (e.g., see [14–25]). A plethora of techniques have been developed to address cybersecurity risks, although these methodologies and techniques tend to be complex and 'multi(dimensionally)-valued' (e.g., but not limited to [26–39]). Often, security has been defined in terms of integrity (assurance that the information, whether received over a network or at rest in storage is just as it should be without malevolent manipulation); system availability (access to the data by authorised users should be unimpeded and reliable); and confidentiality (intelligibility of the information should to be restricted and protected to authorised parties). In the context of integrity, a blockchain is a database that stores transactions – information or data – in temporal order in a set of computer storage facilities that are, in principle, manipulation-proof to a nefarious entity. These transactions can, in turn, be shared by all users in the system. Information is published or stored as a public ledger that is incapable of being surreptitiously modified. Every node in the ecosystem maintains the same ledger information as all other users or nodes in the network. Thus, BCMs can be used to address forgery, tampering, and software distribution.

While the IT infrastructure of an organisation is already, de facto, vulnerable, the introduction of broadly dispersed, physically accessible, low-complexity devices under the rubric of the IoT invariably leads to a broadening overall attack surface, and if uncontrolled, this surface could afford hackers and intruders an unmonitored backdoor into an organisation's essential information in their IT domain; this has already happened in well-publicised cases. This enlarged attack surface dictates a need for end-to-end security mitigation. Furthermore, the environment engendered by IoT applications can increase the overall risk of experiencing IT infrastructure breaches. Additionally, as noted, many high-density applications – for example in Smart cities' environments – make use of physical gateways that aggregate local traffic; these gateways or fog nodes may represent a 'low-hanging fruit' exploitable for infiltration. For example, in recent years there have been major concerns related to power plant and electric grids infiltration and disruption and also concerns related to gas pipeline grid infiltration and disruption resulting from the potential compromise of supervisory control and data acquisition (SCADA)-based management and control systems. Webcams and other cameras continue to be particularly vulnerable [40].

Besides the fact that many IoT endpoints have limited computing, memory, and power resources, constraining the amount of security-related processing, some of the evolving technologies could further exacerbate the task of securing the IoT ecosystem. For example, 5G cellular systems will likely increase the number and types of cells, ranging from microcells to macrocells to the cloud radio access network (C-RAN). In C-RANs, the baseband processing for many cells is pooled and centralised. C-RANs are expected to embody the concepts of virtualisation, following the planned migration by carriers to network function virtualisation (NFV)-based network elements (NEs). NFV is driven by the need for reduced hardware footprints and power requirements, improved provisioning intervals, and reduced costs. NFV, however, will place additional NEs at risk for possible infiltration and breach, thus requiring some (new) heavy-duty security protection mechanisms.

A number of relatively new security approaches and techniques have been recently proposed or used for IoT environments. At a minimum this could imply some straightforward endpoint firewalling mechanism (although these mechanisms would typically be 'lightweight'), encrypted data streams, firewalling functionality in the fog-backbone handoff (or cloud handoff), firewalling functionality at the data aggregation point, specially designed security middleware, and secure mechanisms in the cloud protecting the analytics engine. There is a need to securely identify endpoint devices, support secure software updates, provide information stream confidentiality, guarantee stream integrity, and for many applications, optimise resource availability [41–47].

As noted, BCMs can play a role in securing a number of IoT-oriented environments by becoming part of a security ecosystem in the context of defences-in-depth approaches [16,48–52]. However, recently hackers have exploited some security lacunae in smart contract and cryptocurrency systems, endeavouring to breach such systems [53]. These lacunae have to be properly assessed and addressed to avoid onerous and expensive future infractions. Blockchain applications in this context are discussed in Sections 2.6 and 2.7.

2.3 IoT ARCHITECTURES WITH EMPHASIS ON THE EDGE STRUCTURES

2.3.1 Architectural concepts

An (IoT) ecosystem is comprised of many functional elements. Preferably these elements can be obtained from a variety of vendors, with assurance that they will interwork. Reference architectures (RAs) are important in this context. Architectures simplify the characterisation of the system's constituent functional blocks and the manner in which these functional blocks interrelate to each other. An RA facilitates the orderly partition of functions, typically in a hierarchical fashion. Such partition not only reduces functional redundancy and also promotes standardisation with the possible definition of well-established layer-to-layer interfaces (also possibly including application programming interfaces [APIs]) but also allows the intermingling of products from an open set of vendors' products with the idea of layer-function cost optimisation or usage of best-in-class technology for each layer. RAs typically embody a number of environment views: A functional view adopts functional-decomposition perspective by specifying interfaces, interactions, and functionality, and an information view adopts an information-structure perspective by defining semantics and information flows. In aggregate, RAs, frameworks, and ensuing standards foster seamless connectivity and plug-and-play operations.

One of the early examples of a RA was the Open Systems Interconnection reference model which has proven instrumental in the development of seamless electronic communications of all types. The IT field also developed a number of RAs, known as enterprise architecture frameworks (EAFs), but because of the multidimensional nature of computing, not a single architecture emerged as the canonical architecture. Some of the architectures that did emerge include, but are not limited to, the Open Group Architecture Framework (TOGAF), the Zachman International model, the US Departure of Defense Architecture Framework (DoDAF), the Federal Enterprise Architecture Framework (FEAF), the NIST Enterprise Architecture Model, the EA^3 Cube Framework, the DND/CF Architecture Framework (DNDAF), the NATO Architecture Framework (NAF), the Unified Modelling Language (UML), and the Decision Model and Notation (DMN), along with those sponsored by vendor or universities such as those promulgated by IBM or Gartner. Some of these RA are also possibly useful in an organisation-wide IoT environment.

In turn, a number of specific IoT RAs and architectural frameworks have emerged of late, each concentrating on some specific abstraction of an IoT ecosystem [54–60]. Some IoT RAs have been developed by standardisation bodies, whereas other architectures have been advanced by vendors. At this juncture there is not yet a broadly accepted IoT RA, although some convergence is anticipated. As is the case in general, several distinct perspectives have been assumed for IoT RAs, for example, a networking/internetworking connectivity perspective, a data-level/data-exchange-level semantics perspective, and a physical-level perspective, namely the hardware or software required to acquire and collect the endpoint information and transfer it upstream. Given the large number of applications in play (also known as 'use cases'), a single RA may not suffice, and the stakeholder may need to work with one or more architectures. Some of the open IoT architectures are shown in Table 2.1. Each of these architectures has strengths and weaknesses [61].

Table 2.1 Partial listing of Key Internet of Things reference architectures

Model/RA	Description
Industrial Internet Reference Architecture (IIRA)	General RA concentrating on functionality domains
Internet of Things Architecture (IoT-A)	RA with a functional and data perspective emphasis
Reference Architecture Model Industries 4.0 (RAMI 4.0)	RA applicable to smart factories
Standard for an Architectural Framework for the IoT	RA advanced by the IEEE P2413 WG with a focus on security and safety
Arrowhead Framework	RA advanced by EU for networked automation of embedded devices
ETSI High level architecture for M2M	Well-known RA but principally focused on M2M, concentrating on networking and communication
IoT RA	RA developed by ISO/IEC WD 30141 seeking to define IoT domains to facilitate interoperability among IoT entities
Open Systems IoT Reference Model (OSiRM)	RA that has a physical hierarchy and allows layer-by-layer security

Abbreviations: EU, European Union; IoT, Internet of Things; M2M, machine to machine; RA, reference architecture.

2.3.2 The open systems IoT reference model

Recently these authors defined a basic seven-layer Open Systems IoT Reference Model (OSiRM) that recognises the fog and edge functionality in an IoT RA environment [62]. The layers of the OSiRM are as follows (also see Figure 2.3):

- *Layer 1*: The 'nearly-unlimited' universe of 'things' that can partake of the automation provided by the IoT. Examples include but are not limited to smartphones, wearables, medical monitoring devices, home appliances, home security devices; homes and buildings mechanical systems (e.g., lighting, HVAC, occupancy); cameras, vehicles, and utility grid elements.
- *Layer 2*: The 'data acquisition' capabilities supported by sensors, embedded microprocessors, embedded electronics, and the corresponding sensor hubs based on various short-range (on location) transmission technologies, whether wireless or wired. The acquired information includes ambiance data, video, and positioning data. Layers 1 and 2 are in a state of symbiosis being that things endowed with sensors become the IoT clients or endpoints of the hubs (and ultimately of the entire system).
- *Layer 3*: The 'fog networking'/'device cloud' mechanism, specifically the neighbourhood area network (NAN), campus area network

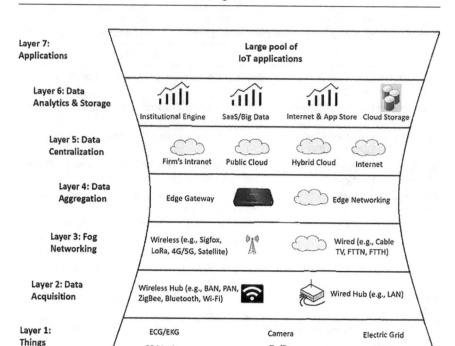

Figure 2.3 Open systems Internet of Things (IoT) reference model (transaction stack).

(CAN), or local site network that is the first link of the IoT device end-to-end connectivity. Fog networking is preferably optimised to the IoT operating environment and may use specialised lightweight protocols. It is often a wireless network (say at the 900 MHz, 2.5 GHz, 5 GHz, mmWave, or infrared/Li-Fi bands) but can also be a wire-based link (e.g., on a factory LAN supporting a process control or robotics application).

- *Layer 4:* This layer supports a(n optional) 'data aggregation' function. It may entail in-network processing, for example, some type of protocol conversion (e.g., converting from a light, low-complexity protocol to a more traditional networking protocol), data summarisation, or other edge-networking functionality related to the outer/access tier of a traditional network, using well-known communication protocols. This function is normally handled in a 'gateway' device.
- *Layer 5:* This is a 'data centralisation' function, corresponding to the classical core networking functions. It encompasses organisation-owned (core) networks, public, private, hybrid cloud-type connectivity, and Internet paths (tunnels). These networks typically consist of carrier-provided telecommunication services and may be wired or wireless.

- *Layer 6*: This layer spans the various 'data analytics and storage functions'. The data analytics functions are typically application specific.
- *Layer 7*: The 'applications' layer spans a large swath of horizontal or vertical application domains (e.g., use cases). In practice, the list of applications is effectively 'unlimited' in scope.

2.4 5G IoT COMMUNICATION TECHNOLOGIES

2.4.1 Wireless technologies for the IoT

Wireless technologies are fundamental to a large number of IoT applications, and as it is tautological, these technologies are intrinsically part of wireless sensor networks (WSNs) [17,41,63]. Wireless access is applicable both in building or campus environments, as well as in city or regional environments (applications also exist for worldwide IoT applications using satellite services of various kinds). Short-distance LAN-oriented technologies, such as Wi-Fi, have been used; city-wide and regional WAN-oriented technologies such as 2G, 3G, and 4G cellular systems (along with other LPWAN services) are also used.

'5G' (5th Generation) is the term for the next-generation cellular and wireless service provider network that aims at delivering higher data rates – 100 times faster data speeds than the current 4G long-term evolution (LTE) technology – lower latency, and highly reliable connectivity. In practical terms, it is an evolution of the previous generations of cellular technology. 5G IoT is licensed cellular-based IoT. A 5G system entails devices connected to a 5G access network, which in turn is connected to a 5G core network. 5G systems subsume important 4G system concepts such as the energy-saving capabilities of narrowband IoT (NB-IoT) radios; secure, low-latency small data transmission for low-power devices – low latency is a requirement for making autonomous vehicles safe; and devices using energy-preserving dormant states when possible. Bandwidth demand for a number of Smart city applications is the main driver for mobile broadband-based 5G services in general and new-generation 5G IoT applications in particular. According to GSMA, 5G is on track to account for 15% (1.4 billion) of global mobile connections by 2025. By early 2019, 11 worldwide operators had announced initial 5G service introductions, and several other operators had activated new base stations with 5G commercial services to follow thereafter [64].

2.4.2 5G wireless technologies for the IoT

Standardisation efforts for 5G systems have been undertaken by several international bodies in recent years with the goal of establishing one unified global standard. Major standardisation bodies included International Telecommunication Union-Radio Communication Sector (ITU-R) and the 3rd Generation Partnership Project (3GPP). The ITU-R has assessed usage scenarios in three classes: ultra-reliable and low-latency communications (URLLC), massive machine-type communications (mMTC), and enhanced mobile broadband (eMBB). Key performance indicators are identified for each of these classes, such as spectrum efficiency, area traffic capacity, connection density, user-experienced data rate, peak data rate, and latency, among others. The ability to efficiently handle device mobility is also critical. Some examples of eMBB use cases include smartphones, home, enterprise, and venues applications, UHD (4K and 8K) broadcast, and virtual reality and augmented reality. eMBB will likely not be immediately used for generic IoT applications because traditional IoT applications tend to be at the lower end of the bandwidth scale. mMTC use cases include smart buildings, logistics, tracking and fleet management, and smart meters. URLLC cases include traffic safety and control, remote surgery, and industrial control. In broad terms, 5G systems are expected to support the following [65]:

- 1000 × higher mobile data volume per area than current systems (downlink peak data rate: 20 Gbps; uplink peak data rate: 10 Gbps);
- 10 × to 100 × higher number of devices than current systems (connection density is 1,000,000 devices per km^2);
- 10 × to 100 × higher user data rate than current systems (downlink 'user experienced data rate': 100 Mbps; uplink 'user experienced data rate': 50 Mbps);
- 10 × longer battery life for low power IoT devices than current systems (up to a 10-year battery life for mMTC); and
- 5 × reduced end-to-end latency than current systems (latency for eMBB is 4 ms, latency for URLLC is 1 ms).

The 5G system expands the 4G environment by adding new radio (NR) capabilities but in a manner that LTE and NR can evolve in complementary ways. The 5G access network may include 3rd Generation Partnership Project (3GPP) radio base stations or a non-3GPP access network. Either the existing traditional LTE evolved packet core (EPC) can be used (as a transition mechanism) to support the 5G NR, or a new 5G core (5GC) can be deployed.

In the near term, '5G IoT' really equates to NB-IoT and LTE-M technologies. In early 2019 there were nearly 100 commercial deployments of LTE-M and NB-IoT worldwide. At this juncture NB-IoT and LTE-M are providing mobile IoT solutions for smart metering, smart logistics, Smart cities' applications, but only in small deployments to date (as of 2018, there were in the range of 50 commercial NB-IoT and LTE-M networks worldwide [66]).

NB-IoT is a licensed low-power LPWAN technology designed to coexist with existing LTE specifications and providing cellular-level QoS connectivity for IoT devices. NB-IoT was standardised by 3GPP in LTE Release 13. Worldwide geographies with GSM deployments will likely offer NB-IoT in the short term. NB-IoT has support from Qualcomm, Ericsson, and Huawei, among numerous other vendors and service providers. NB-IoT is based on a Direct Sequence Spread Spectrum (DSSS) modulation in a 200-kHz channel. There are several underused 200-kHz GSM spectrum channels, as well as other possible bands such as guard bands. NB-IoT is intended as an alternative to long range (LoRa) and Sigfox. NB-IoT service goals include: (i) low complexity end nodes, (ii) device cost less than $5, (iii) a device battery life expected to last for 10 years if it transmits 200 bytes of data per day, and (iv) uplink latency less than 10s (thus not a true real-time service). NB-IoT operates on 900- to 1800-MHz frequency bands with coverage of up to about 20 miles; it supports data rates of up to 250 Kbps in the uplink and 230 Kbps in the downlink. See Table 2.2 for a comparison of metro-level wireless technologies applicable to the IoT.

LTE-M supports low nodal complexity, low nodal power consumption, high deployment density, low latency and extended geographic coverage, while allowing service operators the reuse of the LTE-installed base. LTE-M is power-efficient system, with two innovations support battery efficiency: LTE extended discontinuous reception (eDRX) and LTE power-saving mode (PSM). LTE-M allows the upload of 10 bytes of data a day (LTE-M messages are fairly short compared to NB-IoT messages) but also allows access to Mbps rates. Therefore, LTE-M can support several use cases. In the United States, major carriers such as Verizon and AT&T offer LTE-M services (Verizon has also announced support for NB-IoT – T-Mobile and Sprint appears to lean in the NB-IoT direction).

Some wireless technologies are better positioned for edge and fog networking that others. For example, LoRa and Sigfox tend to be more citywide and county- or regionwide technologies, possibly with a few centralised antennas serving a wide area; signals are received at these dispersed somewhat remote antennas and then channelled directly to the Internet [67]. Other wireless technologies allow a more hierarchical network tiering where data is aggregated (and possibly summarised) at various points along the way; LTE-M and NB-IoT can operate in both modes, depending on the implementation. Low-power radio systems tend to be more matched to fog- and edge-based network designs.

Table 2.2 Comparison of key fog and edge and core wireless technologies applicable to the IoT

IoT technology	Basic features
NB-IoT	• Mostly edge but can also be a small-geography core • Outdoor usability: up to about 20 miles • Several bands, licensed spectrum • Falls under the 5G 'umbrella'; LTE-based • 0.1–0.2 Mbps data rates, battery ~10+ years • Low cost, low modem complexity, low power, energy-saving mechanisms (high battery life) • Does not require a gateway: Sensor data is sent directly to the destination server (other IoT systems typically have gateways that aggregate sensor data, which then communicate with the destination server) • Reasonable building penetration (improved indoor coverage) • High number of low throughput devices (up to 150,000 devices per cell)
LTE-M Rel 13 (Cat M1/Cat M)	• Mostly edge but can also be a small-geography core • Outdoor usability: up to about 20 miles • Cellular network architecture, LTE compatible, easy to deploy, new cellular antennas not required • Falls under the 5G 'umbrella'; uses 4G-LTE bands below 1 GHz, licensed spectrum • Considered the second generation of LTE chips aimed at IoT applications • Caps maximum system bandwidth at 1.4 MHz • Cost-effective for LPWAN applications where only a small amount of data is transferred (e.g., smart metering) • 1 Mbps upload/download, battery ~10 years • Relatively low complexity and low power modem • Can be used for tracking moving objects (location services provided through cell tower mechanisms)
LoRa	• End-to-end (edge and core) • Does not fall under the 5G 'umbrella' • Outdoor usability: 6–15 miles with LOS • Band below 1 GHz • IoT-focused from the get-go • Proprietary • Low power
Sigfox	• End-to-end (edge and core) • Does not fall under the 5G 'umbrella' • Outdoor usability: 30 miles in rural environments; 1–6 miles in city environments • Band below 1 GHz • Narrowband • Low power • Proprietary • Star topology

Abbreviations: IoT, Internet of Things; LoRa, long range; LTE, long-term evolution; LTE-M, long-term evolution machine type communications; NB, narrow band; LOS, line-of-sight; LPWAN, low-power wide-area network.

2.5 SMART CITY CONCEPTS AND COMMON SMART CITY APPLICATIONS

2.5.1 Urban and city trends

Currently 55% of the global population lives in urban areas and 68% of the global population is projected by the United Nations to be living in urban areas by 2050 [68] (and as a point of observation, the world population has increased sevenfold over the past 100 years). In the United States 83.9% of the population is urban (276,062,331 people in 2019, out of population of 329,093,110) [69]. Given this expected increase cities' populations – with cities already often being under constrained resources, both physically and financially – there is an urgent need to consider IT-based resource management and optimisation in general and IoT-based solutions in particular. The Smart city market is expected to encompass $158 billion in expenditures in 2022, making this market an attractive target for technology developers in general and IoT developers in particular [70]. From the perspective of 2020, some believe that the trend toward increased urbanisation may be reversed in the short term due to Covid-19 and similar pandemics; the jury, however, is still out on this matter.

There have been recent societal changes, pre-Covid-19, especially driven by the 'millennial' generation, where people seek to live in cities rather than moving to (more) distant suburbs. This has been driven by the desire to cut the commute time (which can easily add up to 500+ hours a year) and partake of the cultural amenities of the city (e.g., theatre, ballet, opera, colloquia). People now desire to be able to walk more, use bicycles, or avail themselves of public transportation. This has a number of consequences for the tenure of the city and city services.

2.5.2 Smart city applications of the IoT

Table 2.3 enumerates several key Smart city IoT-based applications and correlates these applications to technical requirements and 5G support, as well as a correlating the applications to broad security requirements and possible BCM applicability. Figure 2.4 depicts graphically some key Smart city IoT-supported applications – note, however, that multiple distinct applications can coexist simultaneously in each edge network. Typical Smart city services include, but are not limited to, infrastructure and real estate management; physical security; power and other city-supporting utilities; electric and other utility manhole monitoring; traffic, transportation, and mobility; pollution monitoring; environmental monitoring; flood abatement; livability and smart services; logistics; and Smart city lighting. In the context of Table 2.3, the following are guidelines for bandwidth: 'low bandwidth': 200 kbps or less; 'medium bandwidth': 200 kbps to 2 Mbps; 'high bandwidth': more than 2 Mbps. The following are guidelines for latency: 'low latency': real-time; 'medium latency': 1–5 seconds. The following are guidelines for reliability: 'high reliability': 99.999% availability; 'medium reliability': 99.99%

Table 2.3 Key Smart city challenges and IoT, 5G, and BCM support

Smart city issue and requirements	IoT support/solutions	5G applicability	Bandwidth needs	Latency	Reliability	Security	BCM usability
Infrastructure and real estate management. Requirement: monitor status and occupancy of spaces, buildings, roads, bridges, tunnels, railroad crossings, and street signals.	Networked sensors to provide real-time and historical trending data allowing city agencies to provide enhanced visibility into the performance of resources, facilitating environmental and safety sensing, smart parking and smart parking meters, smart electric meters, and smart building functionality.	High	Low	Low	Medium-to-High	Medium	Medium
Physical security. Requirement: security in streets, parks, stations, tunnels, bridges, trains, buses, ferries.	Networked sensors (possibly including drones and gunshot detection systems) to support surveillance video, license plate reading, gunshot detection, biohazard and radiological contamination monitoring, face recognition, and crowd monitoring and control.	High	High	Low	High	High	High
Power and other city-supporting utilities. Requirement: Reliable flow of electric energy, gas, and water; optimised waste-management and sewer; safe storage of gasoline.	Smart grid solutions and sensor-rich utility infrastructure	High	Low	Medium	High	High	High
Electric and other utility manhole monitoring. Requirement: Electric power manholes require monitoring to avoid or prevent dangerous situations.	Cost-effective and reliable sensors are needed. Technology being deployed by Con Edison in New York City	High	Low	Medium	High	Medium	Medium

(Continued)

Table 2.3 (Continued) Key Smart city challenges and IoT, 5G, and BCM support

Smart city issue and requirements	IoT support/solutions	5G applicability	Bandwidth needs	Latency	Reliability	Security	BCM usability
Traffic, transportation and mobility. Requirement: Optimised traffic flow, low congestion, low latency and high expediency, low noise, minimal waste of fuel and carbon dioxide emissions, safety.	Networked sensors to support traffic flow, driverless vehicles including driverless bus transit, and multi-modal transportation systems. For driverless vehicles, sensors will allow high-resolution mapping, telemetry data, traffic, and hazard avoidance mechanisms.	High	Medium-to-High	Low	High	Medium-to-High	Medium
Pollution monitoring. Requirement: Monitor emission of dioxins, vapourised mercury, nanoparticles, radiation from factories, incinerators, urban crematoria, especially if these sources are close to train tracks or other wind-turbulence elements (e.g., canyons).	Networked sensors throughout town (or within 10 km of a point source) to monitor toxic, health-impacting emissions from point sources including factories, generation plants (if any) and crematoria (if any).	High	Medium	Medium	High	Medium	Medium
Environmental Monitoring. Requirements: Monitor outdoor temperature, humidity and other environmental gases.	Sensors that can be placed in easy-to-deploy locations (e.g., atop existing Smart city light poles to continuously monitor temperature, humidity, and other environmental gases).	High	Low	Medium	Medium	Medium	Medium
Flood Abatement. Requirement: Flood and storm drainage control.	Distributed ruggedised sensors to monitor flood and storm drainage to provide early warning and fault detection.	High	Low	Medium	High	Medium	Medium

(Continued)

Table 2.3 (Continued) Key Smart city challenges and IoT, 5G, and BCM support

Smart city issue and requirements	IoT support/solutions	5G applicability	Bandwidth needs	Latency	Reliability	Security	BCM usability
Livability and Smart Services. Requirement: Quality of life, expeditious access to smart services, efficient transportation, low delays, safety.	Networked sensors (possibly including drones) to facilitate smart multi-modal transportation, information-rich environments, smart parking, location-based services, real-time connectivity to health-monitoring resources (e.g., air quality).	High	Medium	Medium	Medium	Medium	Medium
Logistics. Requirement: Supplying city dwellers with fresh food, supplies, goods, and other materials.	Networked sensors (possibly including drones) to enable the streamlining of warehousing, transportation, and distribution of goods. Traffic management is a facet of such logistical support.	High	Medium	Medium	High	Medium-to-High	Medium-to-High
Smart City Lighting. Requirement: Conversion to LED lighting and associated control via IoT for weather conditions, phases of the moon, seasons, traffic occupancy, and so on.	Cities spend large amounts of money yearly for street lighting (usually 1000 street lights per 10,000 inhabitants). LED lighting requires one-third the amount of power for the same amount of luminance. Sensors are needed for IoT-directed light management for weather conditions, phases of the moon, seasons, traffic occupancy, and so on.	High	Medium	Medium	Medium	Medium	Medium

Abbreviations: BCM, blockchain mechanisms; IoT, Internet of Things; LED, Light Emitting Diodes.

Figure 2.4 Illustrative examples key Smart city Internet of Things (IoT)-supported applications.

availability. The following are guidelines for security: 'high security' means impervious to infractions by state-sponsored hacking activities; 'medium security' means impervious to infractions by professional 'black hat' hackers. The following are guidelines for BCM-mechanisms: 'High usability' means a strong technical and financial advantage to the use of such technology; 'medium usability' means a reasonable technical and financial advantage to the use of such technology. See Figure 2.5.

2.5.3 Formal definition of a Smart city

When researchers endeavour to rank cities on a Smart city index scale, parameters are often evaluated, including social cohesion, human capital, economy, governance, mobility and transportation, environment, urban planning, international outreach, and technology [71]. IoT principles and also BCM mechanisms support all these facets. For example, the 2019 IESE Cities in Motion Index (CIMI) ranks New York, London,

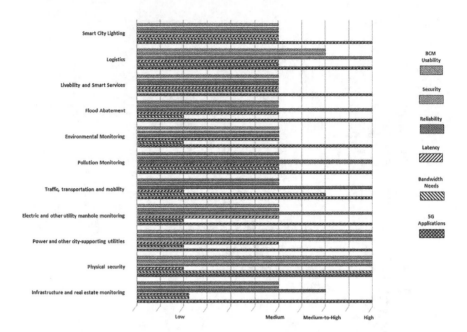

Figure 2.5 Smart city requirements, also including security.

Paris, Tokyo, Reykjavik, Singapore, Seoul, Toronto, Hong Kong, and Amsterdam among the 'list of smartest cities'. More formally, the International Organization for Standardization (ISO) has published a standard, ISO 37120-2018, 'Sustainable Development Of Communities – Indicators For City Services And Quality Of Life (QoL)' (originally issued in 2014 as ISO 37120-2014) that defines and delineates methodologies for a set of standardised indicators to guide and measure the performance of city services and QoL; this set of indicators provides a uniform paradigm for what is measured and how that measurement is to be undertaken and establishes a common basis for reporting, comparison, and benchmarking; it facilitates an integrated approach to sustainable Smart city governance and development (ISO requires third-party verification of data sources and quality to secure ISO certification). ISO 37120 is applicable to any city, municipality, or local government. The standard defines key city themes of city services and QoL. It defines 19 sectors and services provided by a city: economy, education, energy, environment and climate change, finance, governance, health, housing, population and social conditions, recreation, safety, solid waste, sports and culture, telecommunication, transportation, urban and local agriculture and food security, urban planning, wastewater, and water. See Figure 2.6. For each of these sectors, the standard gives a key indicator that should be reported by administrators implementing the standard. ISO 37120-2018 also identifies a profile indicator for each and a

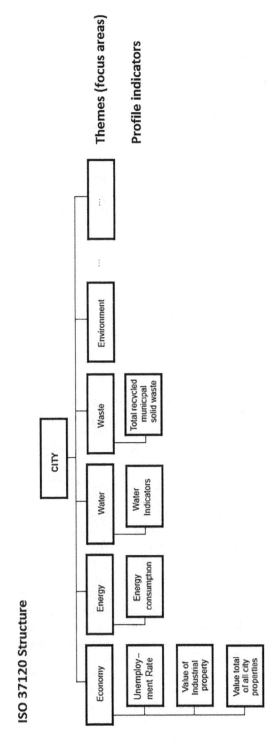

Figure 2.6 ISO 37120 schema.

set of supporting indicators [72]. For example, for the economy sector, the core indicator is the city's unemployment rate [73]. Again, IoT principles and also BCM mechanisms support all these facets.

Tools to implement the ISO 37120 specification typically encompass three modules [74]:

- City model module covering all major urban processes and typical infrastructure as defined in the standard for the schematic themes in conjunction with indicators;
- City data module that allows responsible city departments and organisations to generate monthly or weekly files of data consistent with the standard's requirements; and,
- Smart city monitor module which is typically a high-level AI-driven IoT technology platform for automatically collecting big data from a large number of data sources and mapping this data into actionable customised information.

2.6 SUMMARY OF APPLICABLE BLOCKCHAIN CONCEPTS

2.6.1 Basic blockchain concepts

As discussed elsewhere in this book, blockchains support trust in data information systems without requiring institutionalised third-party verification or audit mechanisms. Blockchains enable participants to achieve unanimity in a distributed network sans central trust authorities.

A blockchain is a cryptographically protected database maintained by a network – specifically, a peer-to-peer (P2P) network – of hosts called 'nodes'. Nodes in the network are typically (but not always) anonymous entities (processes, individuals, or users). Each host (computer, peer) stores a copy of the most up-to-date version of the database. A blockchain block is collection of valid transactions; a blockchain is a cryptographically linked group of blocks generated by nodes. Each block is comprised of a header, the pertinent data to be protected, and related security data (e.g., creator identity, creator signature, last block number, hash of the prior block in the blockchain). The linked blocks form a chain. A blockchain protocol (a consensus mechanism) is a set of rules that codify how the hosts in the network can verify new transactions and add them to the database [75]; the protocol uses cryptographic mechanisms (especially hashing), game theory principles, and incentives to create motivations for the nodes to work toward securing the data in the network, instead of compromising it for individual gain (e.g., see [76–86]). Various nodes perform various functions depending on their role. To support and make use of the blockchain, peer nodes must support a number of key functions, including routing, storage, wallet services, and mining (see Figure 2.7).

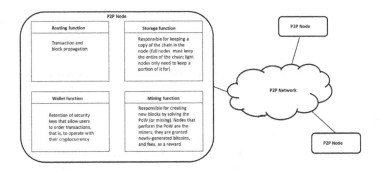

Figure 2.7 Node functions.

'Miners' are nodes that can add or validate transactions; transaction validation (mining) is a fundamental element of the distributed consensus protocol:
nodes in the P2P network are required to validate each transaction of each
block. The proof-of-work (PoW) needed for transaction validation is comprised of a computationally complex algorithm that must be executed for the
generation of blocks; this work activity must be, by design, a computationally
demanding task to carry out to solve and, simultaneously, be easily verifiable once done. Typically, the consensus mechanism includes three phases: a
transaction endorsement process; an ordering process; and a validation and
commitment process. After a miner completes the PoW – executing the PoW
algorithm is an activity also known as 'mining' – the miner publishes the new
block in the P2P network; other nodes in the P2P network verify the validity
of the block before adding the block to the chain they themselves store.

2.6.2 Data integrity using blockchains

Blockchains thus facilitate 'decentralised consensus' by being a distributed
'open-platform' ledger (database) that retains a (developing) list of records,
while at the same time precluding revisions of such records after the fact.
Blockchains provide data integrity across multiple parties by providing
all users in the ecosystem with proof of disintermediated trust; traditionally, this assurance of integrity had to be accomplished by an independent
trusted third party. The blockchain provides openness, incorruptibility,
accessibility, and the ability to store and transfer data in a secure manner by design; it offers a mechanism for entities that do not know or trust
each other to generate a generic shared record or a shared record of asset
ownership. The information held on a blockchain is a shared database that
is consistently and automatically reconciled; the database is not stored in
any discrete or single host or computer but instead the data is stored as
groupings (blocks) of information that are identical in all nodes in the P2P
network; therefore, the blockchain cannot in principle be deleted, altered,
or controlled by any individual node. Blockchains can be public, private, or

consortium: Public blockchains have ledgers visible to everyone on the network and anyone can verify and add a block of transactions to the blockchain (e.g., a cryptocurrency); private blockchains only allow specific nodes in the private organisation to verify and add a transaction block, however, everyone on the network can typically view the chain; in consortium blockchains, only a set of organisation (such as financial institutions) can verify and add transactions, however the ledger can be visible to all or can be restricted to a specified community [87,88].

Thus, nodes can create new blocks of transactions containing various types of data (to be protected); they can validate and digitally sign the transaction; and they can undertake 'mining' to achieve consensus using the consensus protocol (other ancillary functions can also be supported by the nodes). Any node can typically generate a transaction and broadcast it to all nodes in the P2P network. The network nodes validate the transaction using a consensus algorithm that employs the extant transactions. After the transaction is verified, the transaction and the related metadata is combined with other transactions to generate a new block which is added to the existing blockchain. When a transaction is generated by a node it is placed into a local pool of unconfirmed transactions. Miners in the network gather and select transactions from said pools to form a block of transactions. These other miners endeavour to verify the transaction, and if they can individually validate it, the block is added to their copy blockchain (the miners reach consensus). To add this transaction to the blockchain, miners need to find a solution to a complex computationally intensive mathematical problem by executing the PoW algorithm. When a miner finds a solution, the solution is broadcasted (along with their block) to the other miners on the network; the other miners are able to verify it, if all transactions inside the block are valid, consistent with the existing record of transactions on the blockchain [76]. Figure 2.8 summarises the key high-level activities of miners. The number of transactions in a block correlates with

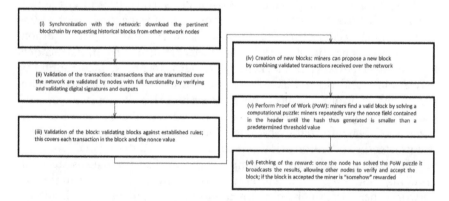

Figure 2.8 High-level activities of miners.

the computational effort required to run the consensus protocol; in turn, this computational effort has a direct impact on confirmation time – transaction validity or lack thereof. It follows that the consensus protocol has a major impact on the scalability of the blockchain-based environment.

If implemented correctly, this blockchain arrangement can make it expensive and computationally difficult to add fraudulent transactions, but at the same time makes it relatively easy to verify that a given transaction is valid. Because blockchains are in principle resistant to modification of the underlying data, they are being considered to support registers of transactions covering anything of value; the information in the blockchain can represent a broad variety of facts, documents, financial transactions, agreements, contracts, signatures, textual elements, or simply data packets [50–52,89–91]. In fact, a number of financial firms, stock exchanges, and even some central banks are planning to use blockchains as part of their operations. The original application of blockchains was for the Bitcoin cryptocurrency; there are now in excess of 1,000 cryptocurrencies in the market.

In summary, the integrity of the information contained in blockchain is protected by the consensus mechanism, and the longest branch of blocks is the one that is considered to be the valid one. The PoW is designed in such a manner that it is computationally complex (pricey) for a nefarious agent to modify a block: Other P2P miners are in the position to invalidate the nefarious agent because the trusted collection of blocks acts to discredit the block that might be generated by a nefarious agent; for a compromised block to be successfully added to the chain, it would require the nefarious agent to solve the PoW faster than the rest of the nodes in the P2P network, but this is computationally challenging, requiring to have control of at least 51% of the computing resources in the P2P network.

2.6.3 Possible weaknesses of blockchains

Although blockchains are generally considered 'unhackable', there are subtle weaknesses, although vulnerabilities may relate more to the implementation than the intrinsic blockchain concept itself. The more complex a blockchain system is, the higher the chance the practitioner implementing the system will invariably make some mistake while setting it up. To give an example, while it may be very difficult to decrypt a stream, it may be possible, instead, to place a trojan or malware in the device undertaking the encryption and thus expose the key or the algorithm being used. Specifically, 'subtle cryptographic flaws' unwittingly built into a cryptocurrency protocol have been documented; leaky or ill-designed client (node) software is also a potential risk.

Many commercial blockchain implementations use a PoW process as the canonical protocol for verifying transactions. As noted previously, the protocol is formulated on the concept that the majority (e.g., 51% or some higher threshold) of the participants (miners) on the network decide which version of the blockchain represents the authoritative chain (database). Nodes need to invest large amounts of computing power to demonstrate they are trustworthy peers are able to add information about new transactions to the database. However, a miner who might surreptitiously gain control of a large amount, specifically a majority of the network's PoW power, can confabulate and defraud other nodes by manufacturing an alternative version, called a 'fork', of the blockchain. A nefarious agent that is able to control the majority of the mining power can institute the fork as being the authoritative version of the blockchain in question. This is known as the '51% attack'; these have appeared recently in the context of cryptocurrencies because by performing a 51% attack on the cryptocurrency blockchain, one can reverse a cryptocurrency transfer, and double-spend the currency. These recent attacks have targeted cryptocurrencies because of the financial gains to the perpetrator – reportedly, hackers have already stolen about $2 B (less so for other applications as of yet); however, the expectation is that 51% attacks will grow in frequency and severity in the future [43]. ASIC mining is another threat; it enables mining companies to enhance the raw power of the mining hardware, greatly enhancing its computing power. Smart contracts based on blockchains are being created in various industries [92]; these contracts automate the movement, execution, and filing of financial-oriented activities (transfers, sales, leases, and so on) and other activities (e.g., voting, online gambling), consistent with defined rules and conditions. Unfortunately, there have been documented infractions based on flaws in the software, giving rise to difficult predicaments when endeavouring to rectify the matter and quell propagation issues. Another issue is exposure to Sybil attacks, which is an attempt to control a P2P network by creating multiple fake identities; although these fake identities appear to be unique users to outside observers, a single nefarious entity controls many identities at once, allowing the nefarious entity to influence the network through augmented voting power in a democratic network (or via echo chamber messaging in a social network) [93].

2.7 SURVEY OF SMART CITY IoT APPLICATIONS THAT ARE OR WILL BENEFIT FROM BLOCKCHAINS

2.7.1 Sample IoT blockchain applications

There are relatively many applications of blockchains documented in the literature: these may or may not be in the IoT context per se, but rather be 'generic' in nature.

'Generic' applications include, but are not limited to financial transactions (e.g., micro-payments, [94–99]), mobile commerce in an IoT environment, healthcare and medical systems (e.g., [100–106]), smart contracts, smart logistics (e.g., verification of the authenticity of items through multi-stage or multi-national supply, tracking controlled items such as pharmaceuticals, medical devices, arms, negotiable bonds, [107]), insurance claims filing and processing (e.g., to monitor or prevent claim fraud detection), smart services (e.g., parking, real-time bus or train arrival monitoring), and cybersecurity management (e.g., data integrity). Several traceability systems and applications have been documented in the literature. For example, 3TG minerals (tantalum, tin, tungsten, and gold), important in the high-tech industry, may (occasionally) have suspicious origins; a number of large companies are now starting to use blockchain mechanisms to facilitate the tracking of 3TG metals and other commodities [108].

IBM has been a key advocate of the IoT and blockchain technology in recent years; Walmart and IBM tested a food supply chain project 2019, and Maersk Shipping is already using blockchains in its shipping process [109,110]. More generally, reference [111] lists 41 documented blockchain applications in the following areas: cryptocurrency, new payment infrastructures, identity verification, e-government, verification of ownership or provenance, product traceability, e-health, energy, insurance and mortgages, real estate (renting, sharing, and selling), gambling and betting, cloud storage, education, and music royalty payments.

Blockchains can in principle be used at the application level of the RA of an IoT architecture to validate all kinds of transactions; essentially, the IoT can use blockchains to safeguard integrity of the business logic data [112]. However, not all IoT applications and not all IoT Smart city applications can (cost)effectively make use of blockchains at this juncture.

2.7.2 Sample IoT blockchain applications for Smart cities

In the Smart cities' environment, the IoT is characterised by having a highly distributed and decentralised physical footprint. Smart cities blockchain applications have, in fact, emerged (as described in, but not limited to, [16,113–118]). Some of the Smart city applications (but not all) are IoT-based. There are forays into two-wheel mobility for city life, including e-bikes with motors and batteries; these vehicles may well be outfitted with IoT sensors and possibly blockchain technology. Reference [111] also lists 11 IoT blockchain applications, some of which may have Smart city applicability: energy microgrid; smart contracts involving IoT devices; renting, selling, or sharing smart IoT objects; insurance for IoT assets; investment in IoT devices; sharing airspace for IoT-based drone navigation; identity, security, and interoperability; data provenance and automation; data integrity for the supply chain; sharing and machine economy; and secure connectivity between IoT devices.

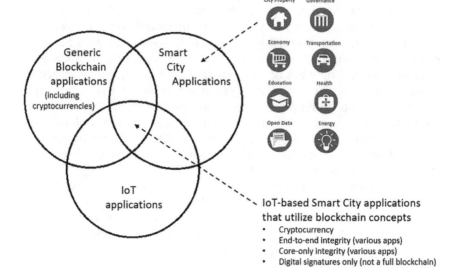

City Property Governance

Economy Transportation

Education Health

Open Data Energy

IoT-based Smart City applications
that utilize blockchain concepts
- Cryptocurrency
- End-to-end integrity (various apps)
- Core-only integrity (various apps)
- Digital signatures only (not a full blockchain)

Figure 2.9 Venn diagram of Internet of Things (IoT) applications.

Figure 2.9 provides a graphical view of the overlapping categorisation of the applications. IoT-oriented Smart city blockchain-based applications may be categorised as follows (this chapter focuses on items 2 and 3):

1. Cryptocurrencies (for various payments of mobile commerce transactions);
2. End-to-end integrity, for example logistics and supply chain, surveillance and 'ring of steel', V2V (vehicle-to-vehicle) and V2I (vehicle-to-infrastructure) traffic, smart grids, e-health, insurance and banking, security, and IoT device management. In some cases, this could also include some encryption;
3. Core-only integrity, for example, municipal smart services and insurance. In some cases, this could also include some encryption; and,
4. Digital-signature-only protection of the data flow.

Some proponents have endeavoured to develop a taxonomy of IoT services or BCM-supported IoT services. Given the rapidly evolving nature of the IoT space any such effort is doomed to short-term obsolescence; nonetheless, Figure 2.10 provides one such evanescent provisional taxonomy.

Platforms are environments upon which IoT or blockchain-based IoT applications, distributed applications, projects, or prototypes can be developed. Open-source platforms are preferable. One well-known platform (among many) is IBM's Blockchain platform; Ethereum is another commonly used platform for IoT-blockchain applications. Platforms typically

Figure 2.10 Example of an exercise aimed at establishing a taxonomy for Internet of Things (IoT) services.

provide various components for consensus and membership, security, identity, and coordination and privacy; they may also include smart contract capabilities.

2.7.3 IoT security blockchain applications

In the context of IoT, blockchains can be part of an arsenal of security measures, particularly for integrity – integrity, which is one of the key security requirements of a transaction, the other requirements typically being confidentiality, authentication, authorisation, and (service/asset) availability. To increase the level of confidentiality and privacy, data in the blockchain can be encrypted [50]. However, certain characteristics of IoT sensors impose several limitations of the security measures that can be employed, employed cost-effectively, or employed in a manner that makes the approach scalable. The low complexity of the device and the low available on-board energy limit the type of computing that can be done at the source point, such as computing for encryption, firewalling, or PoW activities. The limited resources associated with IoT devices often make them unsuitable for directly supporting consensus mechanisms and PoW. A well-known rule is security is that the value of the data or asset being protected must exceed the cost of the security measure, preferably by a significant amount. Some IoT data flows do not necessarily require high integrity because the data might be perishable (only have instantaneous or near-instantaneous value), be quickly replaced by updated data, or just represent some incidental monitoring. Examples in this category in a Smart city context might include data generated by environmental monitoring (e.g., outdoor temperature, humidity, pollen counts, etc.); data regarding real estate management (e.g., monitor status and occupancy of spaces, buildings, roads, bridges, tunnels, etc.); or data related to livability (access to city events, general webcams, and train or bus schedules and next arrival). Some IoT data flows may require high integrity, chain-of-command, and assurance of incorruptibility. Examples in this category in a Smart city context might include data related to physical security (e.g., surveillance of streets, parks, stations, bridges); or data related to power and other city-supporting utilities (e.g., smart grid, reliable availability of electric energy, gas, and water; optimised waste-management and sewer; safe storage of dangerous chemicals, etc.) Blockchains would be more ideally used in cases where high integrity and data durability are required. Refer back to Table 2.3 for additional examples of applicability.

Specifically, node processing overhead and network latency may be limiting factors because IoT devices often have limited storage, computation, and communication capabilities. Many blockchain mechanisms including support of appropriate P2P protocols, support of the consensus algorithms, PoW and other related mechanisms, and general node and peer functionality entail a level of overall processing complexity, especially if the blockchain mechanisms are implemented over an entire IoT ecosystem. Given the technical limitations of implementing various P2P nodal roles in generic IoT endpoints (limited computing and storage capacity) to manage distributed ledgers, BCMs may have to be supported in selected nodes in the network, such as in gateways and fog-to-core handoff points [119]. Figure 2.11 relates the architecture layer, application type, and thing and node type (or network node [gateway] type) to the applicability of a blockchain solution for integrity. In reality, it would not be reasonable to expect that a low-end outer-tier IoT node, such as remote actuator or sensor, would or should vouch for integrity of the global ecosystem data. Given the typical limitations of IoT nodes, it may not be feasible to use BCM-based integrity in the global IoT context, but specific critical or mission-critical applications such as smart grids, ITSs, and certain e-health and insurance data flows may be situated in the context of nodes that have adequate capabilities to support the requisite P2P BCM functionality.

An alternative is to design P2P IoT networks having scope that is locally limited; in this approach the volume of blocks or transactions to be processed and stored will be smaller, and simultaneously, the background P2P

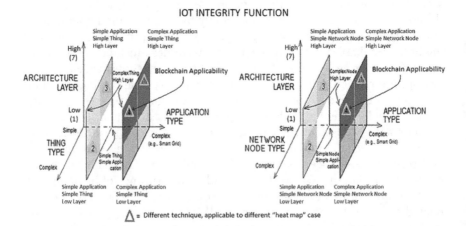

Figure 2.11 Applicability of blockchain solutions.

messages – discovery, querying, invoking, synchronizing, and consensus building – entail less aggregate processing. Yet another alternative is to use a basic ledger in which blocks are digitally signed by the originator or some intermediate node in the fog but without implementing the full consensus process in the fog; the full blockchain mechanism can be implemented as the data transits through nodes closer to or in the core or cloud. Additionally, not all types of data needs absolute integrity because of its perishability.

2.7.4 Blockchain implementation modes in the IoT

It turns out that in the IoT environment, several implementation modes are possible for blockchains. Implementations include end-to-end blockchains, analytics and storage-level blockchains, gateway-level blockchains, site-level blockchains, and device-level blockchains. See Figure 2.12. In the end-to-end approach one would have the end node create BC blocks, but given the relative simplicity of an inexpensive and somewhat 'passive' end node, this design is generally impractical for typical end nodes (although in some specific environments, this may be doable). Another approach is where the end sensor does not create new blocks of transactions but simply passes the data along to the edge gateway or fog node; the edge gateway could then create or add blocks and have the data protected from that point on. It should be noted that fully participating nodes are also required to support P2P messages to deal with (i) discovery of other peers in the blockchain-secured P2P network; (ii) transaction management such

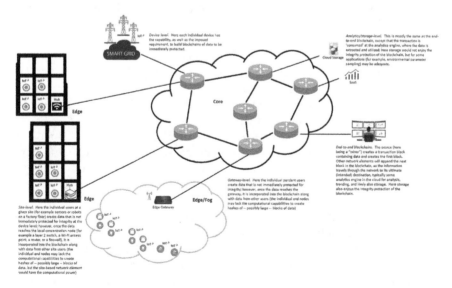

Figure 2.12 Implementation modes for Internet of Things (IoT) blockchains.

as transaction querying, invoking, and deploying; (iii) synchronisation to keep the blockchain updated on all P2P nodes; and (iv) consensus management such as endorsing the transaction. All of these activities and messaging require non-trivial amount of computing, memory storage, and physical power. Not all IoT use cases have nodes that can support these activities.

2.7.5 Specific Smart city blockchain applications

A handful of specific applications are briefly discussed next.

2.7.5.1 E-health

E-health has important (IoT) applications both in hospitals and in the open [100–106]. Hospitals are part of the set of city resources, and large cities have many hospitals, particularly specialised tertiary centres. For example, there are 5,262 community hospitals in the United States; of these 3,387 are in urban settings. Some basic statistics about health care include the following [120–123]:

- There are 77 hospitals within 25 miles of New York City; 51 hospitals in the Chicago area; 68 in the Los Angeles area; compare with 11 hospitals within 50 miles of Peoria, Illinois, and 10 hospitals within 50 miles from the centre of Sioux City, Iowa (source: Medicare).
- On average, there are about 200 beds per hospital (source: American Hospital Directory).
- Average number of beds per 1,000 people in the population is 2.8.
- As an average, a city with 1 million residents would have 2,800 beds or 14 hospitals. A city with 10 million residents would have 28,000 beds or 140 hospitals.
- According to the 2019 Admissions in Community Hospitals: 34,305,620; the national average length of stay (ALOS) is 4.5 days (at an average cost of $10,400 per day).
- Hospitals have been closing at a rate of about 30 a year, according to the American Hospital Association.
- US health care spending grew 3.9% in 2017, reaching $3.5 trillion or $10,739 per person. Health spending accounted for a 17.9% share of the nation's gross domestic product, (source: National Health Expenditure Accounts [NHEA]).

In-hospital applications include patient identification; lab reports management; medication administration and tracking; and medical asset tracking. The health care industry is highly regulated, particularly in the context of patient records. For example, if the average stay in the hospital is 4.5 days, and 10 records or lab results or documents a day are generated, that equates to, say, 50 records per hospital stay, or about 1.7 billion records a year in the

United States only. Just in the United States, the following regulation applies: 42 Code of Federal Regulations (CFR) Part 2; the Federal Information Security Management Act (FISMA); the General Provisions Applicable to Part 164 – the Security, Breach Notification, and Privacy Rules (Part 164, Subpart A); the Genetic Information Nondiscrimination Act (GINA) of 2008; the Health Insurance Portability and Accountability Act of 1996 (HIPAA); Medicaid (Title XIX of the Social Security Act); the Patient Protection and Affordable Care Act; the Proposed HIPAA Regulations Implementing HITECH; the Breach Notification Rule Table (Part 164, Subpart D); the Common Rule; the Enforcement Rule Table (Part 160, Subparts C, D, & E); the Family Educational Rights and Privacy Act (FERPA); the Health Information Technology for Economic and Clinical Health (HITECH) Act; the Privacy Rule Table (Part 164, Subpart E); the Security Rule Table (Part 164, Subpart C). It is, thus, self-evident that record keeping, tracking, storing, and transfer are critical to the industry. Furthermore, the ability to save money in the future hinges on being able to share documents among various hospitals, lab centres, clinics, family doctors, and pharmacies. Additionally, there are many at-home and at-large e-health applications where mobile monitoring devices of the IoT type are used to collect medical status information in real time and have it transmitted over the cloud for assessment and storage. BCM will play an important role in this space in the upcoming years; for example, gateway and fog-node-level or site-level devices can operate as miners and create a blockchain containing the health data to be transferred to the remote medical institution (encryption of the data may typically also be used).

Recently, a number of technologies have been enlisted to fight the COVID-19 pandemic, including artificial intelligence for medical research, 5G systems to increase information transmission speed for work-at-home applications, and BCMs to manage health care and food supply chains. Specifically, BCMs are being used to monitor pandemic material distribution, to track donations, to track relief distribution, for crisis management, and to support contact-tracing while at the same time retaining appropriate anonymity by adhering to mandated healthcare-data regulations. In early 2020, the U.S. Department of Homeland Security (DHS) published guidelines for the coronavirus pandemic, listing blockchain managers in food and agricultural distribution as 'critical infrastructure workers'. Organizations can collaborate using BCMs to research the best way to combat the coronavirus by providing transparent and immutable medical data. BCMs can be used to track the spread of the virus, the number of infected citizens, and the number of recovered citizens. BCMs can also be used to promote transparency during fund-raising and donation campaigns while providing a means to track the origin of funds and detect if those funds are being used for the right purposes. Utilizing BCMs, companies can receive payment through cryptocurrencies without the risk of spreading the virus through cash – for example, the Covir blockchain project (covir. io) partnered with Octopus Robots to fund Biosafety licenses for robots that decontaminate commercial buildings [124,125].

2.7.5.2 Smart grid

Smart grids entail a transmission and distribution (T&D) component – usually in rural areas – and a consumption component usually in populated areas. Cities are critically dependent on electric power (and other utilities) for their survival and sustainability. BCMs can play important roles in this arena, also in the context of tracking environmental parameters and energy source data.

2.7.5.3 Logistics

Logistics and supply chain applications abound. Cities depend critically on the arrival and distribution of goods, food stuff, and other consumables; without said availability, a city would quickly degenerate into chaos. Food traceability using BCM is becoming an important area of early implementation of the IoT/BMC (one example of global- or satellite-based IoT is the management of truck refrigeration while the goods are in transit across a country – relatively stringent regulation has been introduced in this context in the recent past). Traceability is important for many goods, but especially for food. For example, European Union (EU) regulations require food producers to trace and identify all constituent materials used in the production or processing of food products (e.g., feeds used in the farming of beef, pork, fish; movement of goods, and so on). Traceability of food products is a key element of food safety, especially considering that the World Health Organization (WHO) estimates that every year approximately 600 million people worldwide succumb to illness caused by contaminated food and 420,000 die from the consequences [111,126]. A large food company or retail establishment (e.g., Walmart in the United States) typically has thousands of suppliers and millions of customers; in the EU the information must be computerised to comply with regulation. IoT monitoring and BCM-based integrity will play a critical role in this space in the upcoming years. As noted previously, other traceability applications have emerged; ethical sourcing is becoming important for many commodities, including diamonds. In these applications the incorporation of IoT-sensing devices enable the product to be traced and can provide information during its transportation, transfer, and distribution. Documentation management, especially in import and export applications can be greatly facilitated by BCMs. In some instances, traceability can be combined with smart contracts where the legitimate, legal, and reliable execution of the contract is enforced by the blockchain.

2.7.5.4 Surveillance

General surveillance as well as city-wide 'ring of steel' applications can be enhanced by BCM, especially where custody of records is critical for follow-up legal proceedings. For example, the Lower Manhattan Security Initiative is an effort to establish comprehensive surveillance in lower Manhattan. At press time, 18,000 private sector and law enforcement cameras record people and vehicle activities in real time in New York City (see Figure 2.13 – about 2,700 cameras are connected to the New York Police Department's network, some being fixed view while many are Pan-Tilt-Zoom (PTZ) and even 360-degree cameras; the police have access to more than 9,000 cameras just in lower Manhattan). Broad concepts were initially inspired by the 'ring of steel' in London's financial district, developed in the 1990s, where a half-million cameras now monitor public spaces throughout greater London with 1 camera for every 16 people. BCM, IoT, and AI are (or can be) used in a symbiotic manner to enhance security and capture and secure important city data. In addition, the Domain Awareness System (developed in partnership with Microsoft) collects and analyses information from police cameras, traffic cameras, EZ-Pass scanners, radar sensors, license plate readers, and signal-control boxes to facilitate tracking the movements of cars and people, provide surveillance, and regulate traffic lights. New York City's LinkNYC kiosks, which provide free wireless internet, each have two surveillance cameras. Additionally, New York City's police department have facial recognition algorithms to analyse the video collected by the cameras. Software provided by IBM uses AI in its video analytics software's with the ability to search for individuals using traits such as age and ethnicity, in addition to facial recognition software. Combining New York City police and governments access to the network of camera with a large database of faces (e.g., driver's license database), it becomes possible to track citizens throughout a metro area in real time. Surveillance technology has been used across the United States, including, for example, on the 2,369 miles of US Route 1 traversing major cities on the East Coast [127–132]. In China, the Pingan Chengshi (Safe Cities) program claims to have connected 170 million cameras across the nation in 2018 (approximately one surveillance camera for every seven citizens), with another 400 million units to be installed by 2020; reportedly the goal is to be able to identify anyone, anytime, anywhere in China who is in a public setting within three seconds [131]. In cases in which the surveillance data is intended to be used for legal processes, blockchain technology can prove useful, for example, guaranteeing the content of chain-of-custody of graphics, photographs, images, videos, or data.

2.7.5.5 Municipal smart service

Municipal smart services, under the rubric of e-government, can also use BCMs. Applications include filing and retrieving permits, building applications, licenses, birth certificates, payment of municipal fees, and public parking management. Automated renting and tracking of city bicycles is also a documented application. The process of patrolling an area, such as a parking lot or city street to issue tickets to cars parked illegally or that are parked in an expired time limit, is expected to be further automated by IoT means (also possibly in conjunction with BCMs); for example, the US Bureau of Labor Statistics notes that the projected job growth for parking enforcement workers through 2026 is expected to be –35% by 2026 [133].

Figure 2.13 New York City surveillance cameras.

2.7.5.6 Economic development

Economic development applications in support of decentralised, shared economy aimed at allowing citizens to monetise things to create financial wealth are goals that can also be supported by smart contracts [94–99]. In addition to Uber, Lyft, and Airbnb, there are a plethora of opportunities in the digital economy, including sharing applications, peer-to-peer automatic payment, and digital rights management [134,135]. Another example is the enablement of micro payments in support of cyber-oriented transactions.

2.7.5.7 Vehicular applications

For V2V, V2I, and intelligent transportation systems (ITSs), data integrity is invariably required, as is availability (non-repudiation). In vehicular support functions, IoT data that is generated by a set of sensors in a car or in a group of cars (or other vehicles) may be aggregated at a gateway point and may well benefit from BCMs. On the other hand, pure V2I and V2V data would likely remain local, but some specialised or administrative data may well be forwarded to a cloud-based analytics engine. The summarised data can be treated as a transaction that needs to be secured as a blockchain block that is created at that point in the edge and fog network for downstream processing in the cloud, or stored or archived somewhere. Other transportation-related applications might include insurance data, such as for usage-based insurance (UBI) data, data related to fractional ownership of vehicles, or operational or administrative data generated by autonomous vehicles. Additionally, said data could include video surveillance screenshots, medical sensor data, insurance claims, medical claims, and so on. Some see the emergence of autonomous cars as a service, where most cars are sold to ride-sharing companies such as Uber or Lyft. IoT sensors such as radar, LIDAR, UBI, and so on, may well benefit from some blockchain support [136].

2.7.5.8 Insurance and banking

There are also insurance and banking applications. One documented use of IoT mechanisms for insurance is UBI. In this environment, also known as pay-as-you-drive or pay-how-you-drive, the premiums are based on distance travelled, on the time-of-day when driving occurs, on the location where driving occurs, and importantly, on how the driver behaves in terms of speed, acceleration, and lane changes. A device is embedded in the automobile that collects driving data that is sent back to the cloud (either using fog and edge networking or core networking) for analysis and assessment. This data may need the integrity assurance provided by the blockchain mechanism in a general setting, but it might need such assurance when it involves a municipal vehicle, a bus, a school bus, a police car or ambulance, an Uber/Lyft vehicle, a self-driving vehicle, or a commercial vehicle such as a Fedex or UPS truck. In the latter cases, the blockchain might start in the gateway – again because it might be too demanding to start the blockchain in the remote device in the car itself; on the other hand, some high-end vehicles such as a bus might have an IoT device with sufficient computing power and electric power to be actual blockchain nodes in the fog itself.

2.7.5.9 Device management

IoT security and IoT device management also can benefit from BCM. BCMs allows confidential sharing of IoT data set [137–139]. IoT devices need to communicate and synchronise with each other. Using BCMs one can properly control and configure IoT devices (e.g., manage encryption keys) [140]. Secure software pushes to remote IoT devices can also be accomplished with BCM, although other methods are also possible or have been used [141,142].

2.8 FUTURE RESEARCH AREAS

Assertions such as these encapsulate well the theme of this chapter 'One of the largest industry shifts of all time (perhaps the largest) involves IoT and the infrastructure and services that need to be in place to make IoT a reality. In short, by making various objects in our everyday lives tech-compatible, we can capture valuable information that can change the way we live and change the way that companies operate. ... To achieve an IoT world, though, the roll-out of 5G will be important, as will be the adoption of the cloud.' [143]. BCMs play a role in that emerging all-encompassing ecosystem. As discussed in this chapter, blockchains have a number of potential applications in IoT. However, significant processing and storage requirements for blockchains and the desire to have low-cost and high-longevity nodes (from a power perspective) make the adoption of such an integrity scheme challenging at this juncture.

Hence, there are a number of technical areas that will benefit from ongoing research into blockchain applications for IoT applications, including scalability (e.g., storage capacity scalability), nodal processing optimisation, refined assessment of P2P nodal functions (e.g., which nodes are the ideal implementers of the PoW functionality, lightweight consensus mechanisms), related security mechanisms (e.g., confidentiality, anonymity and data privacy, endpoint authentication, trusted execution environments, and service availability in the context of non-repudiation). Design principles to avoid falling prey to the 51% attacks should be further explored, as well as protection against double-spend attacks, race attacks, Sybil attacks, denial-of-service attacks, eclipse attacks, and man-in-the-middle issues, being that many if not most P2P protocols and IoT environments are vulnerable to these kinds of attacks.

Pragmatic limitations of blockchain technology as applied to the IoT environment (e.g., device performance, communication complexity) should be studied and assessed. Cryptocurrency issues related to mobile commerce in an IoT environment are also an area that will likely require theoretical, technical, and practical assessment.

Finally, a research question might be if BCMs can be used to mitigate protocol-based attacks in the IoT environment, for example internal attacks on fog- and edge-based IoT routing in ad hoc networks; protocol-based attacks include wormholes, blackholes where a malicious node intervenes in route discovery to be a part of path, greyholes, selective forwarding, local repair attacks by sending false link information, route cache poisoning, Sybil attacks, sinkholes, hello floods, neighbour attacks with Destination Oriented Directed Acyclic Graph Information Object (DIO) messages, version number alteration of the DIO message, modification of 'no trust levels', information fabrication, Byzantine attacks to create loops, and location spoofing by purporting to be nearest destined node [116,144].

REFERENCES

1. Hassan, Q. Editor, 2018. *Internet of Things A to Z: Technologies and Applications.* IEEE Press/Wiley, Hoboken, NJ.
2. Hassan, Q. Editor, 2018. *Internet of Things: Challenges, Advances and Applications.* CRC Press, Taylor & Francis Group, Boca Raton, FL.
3. Aazam, M. and E.-N. Huh. 2014. Fog computing and smart gateway-based communication for cloud of things. *Proceedings 2nd International Conference on Future Internet of Things and Cloud,* Barcelona, Spain.
4. Alrawais, A., Alhothaily, A., et al. 2017. Fog computing for the Internet of Things: Security and privacy issues. *IEEE Internet Computing,* 21, 34–42.
5. Byers, C. C. 2017. Architectural imperatives for fog computing: Use cases, requirements, and architectural techniques for fog-enabled IoT networks. *IEEE Communications Magazine,* 55, 14–20.
6. Mushunuri, V., Kattepur, A., et al. 2017. Resource optimization in fog-enabled IoT deployments. *Proceedings of the 2017 Second International Conference on Fog and Mobile Edge Computing (FMEC),* Valencia, Spain, May 8–11, 2017, pp. 6–13.
7. Charalampidis, P., Tragos, E., et al. 2017. A Fog-enabled IoT platform for efficient management and data collection. *Proceedings of the 2017 IEEE 22nd International Workshop on Computer Aided Modeling and Design of Communication Links and Networks (CAMAD),* Lund, Sweden, June 19–21, 2017, pp. 1–6.
8. Azimi, I., Anzanpour, A., et al. 2017. HiCH: Hierarchical fog-assisted computing architecture for healthcare IoT. *ACM Transactions on Embedded Computing Systems (TECS),* 16, 174.
9. Baker, T., Asim, M., et al. 2019. A secure fog-based platform for SCADA-based IoT critical infrastructure. *Software: Practice and Experience.* doi:10.1002/spe.2688.

10. Ni, J., Zhang, K., et al. 2018. Securing fog computing for Internet of Things applications: Challenges and solutions. *IEEE Communication Surveys Tutor,* 20, 601–628.
11. Abbas, N., Asim, M., et al. 2019. A Mechanism for securing IoT-enabled applications at the fog layer. *Journal of Sensor and Actuator Networks,* 8, 16.
12. Liang, K., Zhao, L., et al. 2017. An integrated architecture for software defined and virtualized radio access networks with fog computing. *IEEE Networks,* 31, 80–87.
13. Roman, R., Lopez, J., et al. 2018. Mobile edge computing, fog etc.: A survey and analysis of security threats and challenges. *Future Generation Computer Systems,* 78, 680–698.
14. A. Dehghantanha. Editor. 2018. *Handbook of Big Data and IoT Security.* Springer Nature, Cham, Switzerland. Part of Springer Nature. doi:10.1007/978-3-030-10543-3.
15. Zhou, J., Cao, Z., et al. 2017. Security and privacy for cloud-based IoT: Challenges. *IEEE Communications Magazine,* 55(1), 26–33. doi:10.1109/MCOM.2017.1600363CM.
16. Dorri, A., Kanhere, S. S., et al. 2017. Blockchain for IoT Security and Privacy: The Case Study of a Smart Home. *IEEE International Conference on Pervasive Computing and Communications Workshops.*
17. Sedjelmaci, H., Senouci, S. M. and M. Feham. 2012. Intrusion detection framework of cluster-based wireless sensor network. *2012 IEEE Symposium on Computers and Communications (ISCC),* Cappadocia, Turkey, July 1–4, 2012.
18. Chakrabarty, S. and D. W. Engels. 2016. A secure IoT architecture for smart cities. *Consumer Communications & Networking Conference (CCNC), 2016 13th IEEE Annual,* January 9–12, 2016. doi:10.1109/CCNC.2016.7444889.
19. Li, X., Wang, Q., et al. 2019. Enhancing cloud-based IoT security through trustworthy cloud service: An Integration of security and reputation approach. *IEEE Access,* 7. doi:10.1109/ACCESS.2018.2890432.
20. Stergiou, C. Psannis, K. E., et al. 2018. Secure integration of IoT and cloud computing. *Future Generation Computing Systems,* 78, 964–975.
21. Minoli, D. and B. Occhiogrosso. 2018. Blockchain mechanisms for IoT security. *Elsevier IoT Journal,* 1–2(1), 1–13.
22. Minoli, D., Kouns, J. and K. Sohraby. 2017. IoT Security (IoTSec) considerations, requirements, and architectures. *14th Annual IEEE Consumer Communications & Networking Conference (CCNC),* IEEE, Las Vegas, NV, January 8–11, 2017.
23. Minoli, D. and B. Occhiogrosso. 2018. Security considerations for IoT support of e-health applications. In *Internet of Things: Challenges, Advances and Applications,* Q. Hassan, Editor. CRC Press, Taylor & Francis Group, Boca Raton, FL.
24. Arsi, S., Inukollu, V. N. and S. R. Ravuri. 2014. Security issues associated with big data in cloud computing. *IJNSA,* 6, 3.
25. Lai, C., Lu, R., et al. 2015. Toward secure large-scale machine-to-machine communications in 3GPP networks. *IEEE Communications Magazine Supplement,* 53(12), 12–19.

26. U.S. Patent 9129108. 2012. Systems, Methods and computer programs providing impact mitigation of cyber-security failures.
27. U.S. Patent 9118702. 2011. System and method for generating and refining cyber threat intelligence data.
28. U.S. Patent 9094288. 2011. Automated discovery, attribution, analysis, and risk assessment of security threats.
29. U.S. Patent 9086793. 2006. Methods and systems for assessing security risks.
30. U.S. Patent 9083741. 2012. Network defense system and framework for detecting and geolocating botnet cyber attacks.
31. U.S. Patent 9054873. 2009. Compact security device with transaction risk level approval capability.
32. U.S. Patent 9032318. 2006. Widget security.
33. U.S. Patent 9555772. 2017. Embedded security system for environment-controlled transportation containers and method for detecting a security risk for environment-controlled transportation containers.
34. U.S. Patent 8997230. 2013. Hierarchical data security measures for a mobile device.
35. U.S. Patent 8966640. 2014. Security risk aggregation and analysis.
36. U.S. Patent 8955111. 2011. Instruction set adapted for security risk monitoring.
37. U.S. Patent 8914880. 2012. Mechanism to calculate probability of a cyber security incident.
38. U.S. Patent 8856936. 2012. Pervasive, domain and situational-aware, adaptive, automated, and coordinated analysis and control of enterprise-wide computers, networks, and applications for mitigation of business and operational risks and enhancement of cyber security.
39. Application US 62/284,983. 2017. Method for a uniform measure and assessment of an institution's aggregate cyber security risk and of the institution's cybersecurity confidence index. D. Minoli and B. Occhiogrosso Inventors.
40. Shodan. Available at www.shodan.io.
41. Komal, K. and T. Mohasin. 2015. Secure and economical information transmission for clustered wireless sensor network. *International Journal of Science and Research (IJSR)*, 4(12), 1033–1035.
42. Chen, G., Gong, Y., et al. 2015. Physical layer network security in the full duplex relay system. *IEEE Transactions on Information Forensics and Security*, 10(3), 574–583.
43. Esser, B. and T. Kiesling. 2016. Keynote 1: Today's cyber security threats and challenges for telco providers; Keynote 2: Cyber resilience of complex interdependent infrastructures. *10th International Conference on Autonomous Infrastructure, Management and Security (AIMS 2016)*, June 20–June 23, 2016, Munich, Germany. Available at www.aims-conference.org/2016/
44. Kennedy, D. 2014. Bug exposes IP cameras, baby monitors. January 14, 2014. Available at http://krebsonsecurity.com/2014/01/bug-exposes-ip-cameras-baby-monitors/.
45. Notopoulos, K. 2012. Somebody's watching: How a simple exploit lets strangers tap into private security cameras. *The Verge*, February 3, 2012. Available at www.theverge.com/2012/2/3/2767453/trendnet-ip-camera-exploit-4chan.
46. Granjal, J., Monteiro, E. and J. Sá Silva. 2015. Security for the Internet of Things: A survey of existing protocols and open research issues. *IEEE Communication Surveys & Tutorials*, 17(3), 1294ff.

47. Zhou, L. and H.-C. Chao. 2011. Multimedia traffic security architecture for the Internet of Things. *IEEE Network*, 25(3), 35–40. doi:10.1109/MNET.2011.5772059.
48. Gault, M. 2016. Rethinking security for the Internet of Things. Available at http://techcrunch.com/2016/05/06/rethinking-security-for-the-internet-of-things/.
49. Ouaddah, A., Elkalam, A. A. and A. A. Ouahman. 2016. FairAccess: A new blockchain-based access control framework for the Internet of Things. *Security and Communication Networks*. doi:10.1002/sec.1748.
50. Kosba, A., Miller, A., et al. 2016. Hawk: The blockchain model of cryptography and privacy-preserving smart contracts. *Security and Privacy (SP), 2016 IEEE Symposium on*, San Jose, CA, May 22–26, 2016.
51. Pilkington, M. 2016. Blockchain technology: Principles and applications. In *Research Handbook on Digital Transformations*, F. X. Olleros and M. Zhegu Editors. Edward Elger Publishing, Northampton, MA.
52. Tapscott D. and A. Tapscott. 2016. *Blockchain Revolution: How the Technology Behind Bitcoin Is Changing Money*, Penguin Random House LLC, New York.
53. Orcutt, M. 2019. Once hailed as unhackable, blockchains are now getting hacked. *Technology Review*, February 19, 2019. Available at www.technologyreview.com/s/612974/once-hailed-as-unhackable-blockchains-are-now-getting-hacked/
54. Tiburski, R. T., Amaral, L. A., et al. 2015. The importance of a standard security architecture for SOA-based IoT middleware. *IEEE Communications Magazine*.
55. ISO/IEC. 2014. Study Report on IoT Reference Architectures/Frameworks, 2014.
56. Weyrich, M. and C. Ebert. 2016. Reference architectures for the Internet of Things. *IEEE Software*, IEEE Computer Society, pages 112 ff.
57. Cisco Systems. 2014. Cisco white paper: The Internet of Things reference model. Cisco Systems, San Jose, CA.
58. Intel. 2015. White paper: The Intel IoT platform, architecture specification. Santa Clara, CA.
59. Walewski, J. W. Editor. 2013. Internet of Things Architecture (IoT-A), Project Deliverable D1.2–Initial architectural reference model for IoT. Available at https://cocoa.ethz.ch/downloads/2014/01/1360_D1%202_Initial_architectural_reference_model_for_IoT.pdf.
60. ETSI. 2012. ETSI TS 102 690: Machine-to-machine communications (M2M); Functional architecture. October 2011.
61. Bauer, M., Boussard, M., et al. 2013. IoT reference architecture. In *Enabling Things to Talk*, A. Bassi, et al., Editors. doi:10.1007/978-3-642-40403-0_8, 2013.
62. Minoli, D., Occhiogrosso, B., et al. 2017. IoT Security (IoTSec): Mechanisms for e-health and ambient assisted living applications–A big data role. *The Second IEEE/ACM International Workshop on Safe, Energy-Aware, & Reliable Connected Health (SEARCH 2017)* (collocated with CHASE 2017, Conference on Connected Health: Applications, Systems, and Engineering Technologies), Philadelphia, PA, July 17–19, 2017.

63. Sohraby, K. and D. Minoli. 2007. *Wireless Sensor Networks*. Wiley, Hoboken, NJ.
64. GOS World Staff. 2019. Skyworks unveils Sky5 Ultra platform for 5G architecture. February 27, 2019. Available at www.gpsworld.com/skyworks-unveils-sky5-ultra-platform-for-5g-architecture/.
65. ITU-R SG05 Contribution 40. 2017. Minimum requirements related to technical performance for IMT-2020 radio interface(s).
66. GSMA. 2018. Mobile IoT in the 5G future: NB-IoT and LTE-M in the context of 5G. Available at www.gsma.com/iot/wp-content/uploads/2018/05/GSMAIoT_MobileIoT_5G_Future_May2018.pdf.
67. Lin, J., Shen, Z., et al. 2017. Using blockchain technology to build trust in sharing LoRaWAN IoT. *Proceedings of the 2nd International Conference on Crowd Science and Engineering*, Beijing, China, July 6–9, 2017. doi:10.1145/3126973.3126980.
68. United Nations Department of Economic and Social Affairs. 2018. Available at www.un.org/development/desa/en/news/population/2018-revision-of-world-urbanization-prospects.html.
69. Worldometer Staff. US Population. 2019. Available at www.worldometers.info/world-population/us-population/.
70. World Economic Forum Staff. 2019. Smart cities must pay more attention to the people who live in them. April 16, 2019. Available at www.weforum.org/agenda/2019/04/why-smart-cities-should-listen-to-residents.
71. Berrone, P., Ricart, J.-E., et al. 2019. New York, London and Paris firmly established as the smartest cities. *IESE Insight*. Available at www.ieseinsight.com/doc.aspx?id=2124&ar=&idi=2&idioma=2.
72. ISO 37120. 2018. Sustainable development of communities—indicators for city services and quality of life.
73. Kelechava, B. 2018. Indicators for city services and quality of life in sustainable cities and communities in ISO 37120:2018. August 13, 2018. Available at https://blog.ansi.org/2018/08/indicators-sustainable-city-iso-37120-2018/#gref.
74. Pharos Navigator Staff, ISO 37120 standard as template for Smart City development, Available at https://smartcity.pharosnavigator.com/static/content/en/626/.
75. Sankar L. S., Sindhu, M. and M. Sethumadhavan. 2017. Survey of consensus protocols on blockchain applications. *Advanced Computing and Communication Systems (ICACCS), 2017 4th IEEE International Conference*, IEEE.
76. Bashir, I. 2017. *Mastering Blockchain*. Packt Publishing, Birmingham, UK.
77. Atzei, N., Bartoletti, M., et al. 2017. A formal model of bitcoin transactions. Available at https://eprint.iacr.org/2017/1124.pdf.
78. Brünnler, K., Flumini, D. and T. Studer. 2018. A Logic of blockchain updates. In *Logical Foundations of Computer Science*, S. Artemov and A. Nerode, Editors. LFCS 2018. Lecture Notes in Computer Science, vol. 10703. Springer, Cham, Switzerland.
79. Artemov, S. N. 2001. Explicit provability and constructive semantics. *Bulletin of Symbolic Logic*, 7(1), 1–36.
80. Decker. C. and R. Wattenhofer. 2013. Information propagation in the Bitcoin network. *13th IEEE International Conference on Peer-to-Peer Computing*, pp. 1–10.

81. Matsuo, S. I. 2017. How formal analysis and verification add security to blockchain-based systems. *Formal Methods in Computer Aided Design (FMCAD)*, Vienna, Austria, October 2–6, 2017.
82. Garay, J., Kiayias, A. and N. Leonardos. 2015. The Bitcoin backbone protocol: Analysis and applications. *Advances in Cryptology. EUROCRYPT 2015: 34th Annual International Conference on the Theory and Applications of Cryptographic Techniques*, Sofia, Bulgaria, April 26–30, 2015, Proceedings, Part II, pp. 281–310. doi:10.1007/978-3-662-46803-6_10.
83. Dennis, R., Owenson, G. and B. Aziz. 2016. A temporal blockchain: A formal analysis. *International Conference on Collaboration Technologies and Systems (CTS)*, October 31–November 4, 2016, Orlando, FL.
84. Huang, B., Liu, Z., Chen, J. et al. 2017. Behavior pattern clustering in blockchain networks. *Multimedia Tools and Applications*, 76, 20099–20110. doi:10.1007/s11042-017-4396-4.
85. Awan, M. K. and A. Cortesi. 2017. Blockchain Transaction Analysis Using Dominant Sets. In *Computer Information Systems and Industrial Management*, K. Saeed, W. Homenda and R. Chaki. Editors. CISIM 2017. Lecture Notes in Computer Science, Vol. 10244. Springer, Cham, Switzerland.
86. Ouaddah, A., Elkalam, A. A. and A. A. Ouahman. 2017. *Towards a Novel Privacy-Preserving Access Control Model Based on Blockchain Technology in IoT, Cooperation Advances in Information and Communication Technologies*. Springer International Publishing, New York City, 523–533.
87. Xu, L., Shah, N., et al. 2017. Enabling the sharing economy: Privacy respecting contract based on public blockchain. *Proceedings of the ACM Workshop on Blockchain, Cryptocurrencies and Contracts*, ACM.
88. Dinh, T. T. A., Wang, J., et al. 2017. Blockbench: A framework for analyzing private blockchains. *Proceedings of the 2017 ACM International Conference on Management of Data*, ACM, 2017.
89. Wright, A. and P. De Filippi. 2015. Decentralized blockchain technology and the rise of lex cryptographia. March 10, 2015. Available at https://ssrn.com/abstract=2580664.
90. Xu, X., Pautasso, C., et al. 2016. The blockchain as a software connector. *Software Architecture (WICSA), 2016 13th Working IEEE/IFIP Conference on*, Venice, Italy, April 5–8, 2016. doi:10.1109/WICSA.2016.21.
91. Merkle, R. C. 1988. A digital signature based on a conventional encryption function. In *Advances in Cryptology, CRYPTO '87*. Pomerance C., Editor. Lecture Notes in Computer Science. Vol. 293. p. 369. doi:10.1007/3-540-48184-2_32. Springer, Berlin, Germany.
92. Christidis, K. and M. Devetsikiotis. 2016. Blockchains and smart contracts for the Internet of Things. *IEEE Access*, 4, 2292–2303.
93. Garner, B. 2018. What's a Sybil attack and how do blockchains mitigate them? August 31, 2018. Available at https://coincentral.com/sybil-attack-blockchain/.
94. Eyal, I. 2017. Blockchain technology: Transforming libertarian cryptocurrency dreams to finance and banking realities. *Computer*, 50(9), 38–49.
95. Treleaven, P., Brown, R. G., et al. 2017. Blockchain technology in finance. *Computer*, 50(9), 14–17. doi:10.1109/MC.2017.3571047.

96. Nordrum, A. 2017. Wall Street occupies the blockchain-financial firms plan to move trillions in assets to blockchains in 2018. *IEEE Spectrum*, 54(10), 40–45.
97. Mukhopadhyay, U., Skjellum, A., et al. 2016. A brief survey of cryptocurrency systems. *14th IEEE Annual Conference on Privacy, Security and Trust*, IEEE.
98. Oudejans, J., Erkin, Z., et al. 2017. A framework for preventing double-financing using blockchain technology. *Proceedings of the ACM Workshop on Blockchain, Cryptocurrencies and Contracts*, ACM.
99. Samaniego, M. and R. Deters. 2016. Blockchain as a service for IoT. *Internet of Things (iThings) and IEEE Green Computing and Communications (GreenCom) and IEEE Cyber, Physical and Social Computing (CPSCom) and IEEE Smart Data (SmartData), 2016 IEEE International Conference on*, December 15–18, 2016, Chengdu, China.
100. Mettler, M. 2016. Blockchain technology in healthcare: The revolution starts here. *18th IEEE International Conference on e-Health Networking, Applications and Services (Healthcom)*.
101. Angraal, S., Krumholz, H. M., et al. 2017. Blockchain technology: Applications in health care. *Cardiovascular Quality and Outcomes*, 10(9), e003800. doi:10.1161/CIRCOUTCOMES.117.003800.
102. Azaria, A. Ekblaw, A., et al. 2016. Using blockchain for medical data access and permission management. *IEEE International Conference Open and Big Data (OBD)*.
103. Kuo, T.-T., Kim, H.-E., et al. 2017. Blockchain distributed ledger technologies for biomedical and health care applications. *Journal of the American Medical Informatics Association*, 24(6), 1211–1220. doi:10.1093/jamia/ocx068.
104. Zhang, J., Xue, N., et al. 2016. A Secure system for pervasive social network-based healthcare. *IEEE Access*, 4, 9239–9250. doi:10.1109/ACCESS.2016.2645904.
105. Zhao, H. Zhang, Y., et al. 2017. Lightweight backup and efficient recovery scheme for health blockchain keys. *13th IEEE International Symposium on Autonomous Decentralized Systems*.
106. Xia, Q., Sifah, E. B., et al. 2017. Trustless medical data sharing among cloud service providers via blockchain. *IEEE Access*, 5.
107. Kim, H. M. and M. Laskowski. 2018. Toward an ontology-driven blockchain design for supply-chain provenance. *Wiley Online Library*, March 28, 2018. doi:10.1002/isaf.1424.
108. Nicola, S. 2019. Using blockchain to help fight conflict minerals, *Bloomberg*, April 24, 2019. Available at www.bloomberg.com/news/articles/2019-04-24/using-blockchain-to-help-fight-conflict-minerals?srnd=premium.
109. Sun, L. 2018. IBM and Microsoft are upgrading Walmart's digital supply chain. *The Motley Fool*, September 30, 2018. Available at www.fool.com/investing/2018/09/30/ibm-and-microsoft-are-upgrading-walmarts-digital-s.aspx.
110. Li, C. 2018. Maersk—Reinventing the shipping industry using IoT and blockchain. *HBS Digital Initiative*, June 28, 2018. Available at https://medium.com/harvard-business-school-digital-initiative/maersk-reinventing-the-shipping-industry-using-iot-and-blockchain-f84f74fe84f9.
111. Reyna, A., Martín, C., et al. 2018. On blockchain and its integration with IoT: Challenges and opportunities. *Future Generation Computer Systems*, 88, 173–190. doi:10.1016/j.future.2018.05.046.

112. Ferrag, M. A., Derdour, M., et al. 2018. Blockchain technologies for the Internet of Things: Research issues and challenges. *IEEE Internet of Things Journal.* doi:10.1109/JIOT.2018.2882794.
113. Sun, J., Yan, J. and K. Z. K. Zhang. 2016. Blockchain-based sharing services: What blockchain technology can contribute to smart cities. *Financial Innovation.* Springer. doi:10.1186/s40854-016-0040-y.
114. Ibba, S., Pinna, A., et al. 2017. Citysense: Blockchain-oriented smart cities. *XP '17 Proceedings of the XP2017 Scientific Workshops,* Cologne, Germany. May 22–26, 2017. ACM, New York. doi:10.1145/3120459.3120472.
115. Biswas, K. and V. Muthukkumarasamy. 2016. Securing Smart cities using blockchain technology. *IEEE 14th International Conference on Smart City,* Sydney, NSW, Australia, December 12–14, 2016. doi:10.1109/ HPCC-SmartCity-DSS.2016.0198.
116. Tariq, N., Asim, M., et al. 2019. The security of big data in fog-enabled IoT applications including blockchain: A survey. *Sensors,* 19, 1788. doi:10.3390/ s19081788.
117. Junfeng, X., Tang, H., et al. 2019. A survey of blockchain technology applied to Smart cities: Research issues and challenges. *IEEE Communications Surveys & Tutorials.*
118. Ruizhe, Y., Yu, F. R., et al. 2019. Integrated blockchain and edge computing systems: A survey, some research issues and challenges. *IEEE Communications Surveys & Tutorials.*
119. Zhang, H. Xiao, Y., et al. 2017. Computing resource allocation in three-tier IoT fog networks: A joint optimization approach combining Stackelberg game and matching. *IEEE Internet of Things Journal,* 4(5). doi:10.1109/ JIOT.2017.2688925.
120. American Hospital Association. 2019. Fast facts on U.S. hospitals. Available at www.aha.org/statistics/fast-facts-us-hospitals.
121. Medicate.gov. 2019. Hospital results. Available at www.medicare.gov/hospitalcompare/search.html.
122. American Hospital Directory. 2019. Available at www.ahd.com/state_statistics.html.
123. Center for Medicare and Medicaid Services. 2019. Available at www.cms. gov/research-statistics-data-and-systems/statistics-trends-and-reports/ nationalhealthexpenddata/nationalhealthaccountshistorical.html.
124. Oluwatobi, J. 2020. How blockchain technology can help fighting against COVID-19. Cointelegraph, June 7, 2020. Available at https://cointelegraph. com/news/how-blockchain-technology-can-help-fighting-against-covid-19.
125. Banafa, A. 2020. Blockchain technology and COVID-19. bbvaopenmind. com, June 22, 2020. Available at https://www.bbvaopenmind.com/en/ technology/digital-world/blockchain-technology-and-covid-19/.
126. World Health Organization. 2017. Food safety fact sheet. Available at www. who.int/mediacentre/factsheets/fs399/en/.
127. Shepard, S. 2018. New surveillance cameras to bolster security in NYC. *Security Today,* October 29, 2018. Available at https://securitytoday.com/ articles/2018/10/29/new-surveillance-cameras-to-bolster-security-in-nyc. aspx.

128. Alm, D. 2017. Somebody's watching you: AI Weiwei's New York installation explores surveillance in 2017. *Forbes*, June 15, 2017. Available at www. forbes.com/sites/davidalm/2017/06/15/somebodys-watching-you-ai-weiweis-new-york-installation-explores-surveillance-in-2017/#5d7cfbdc4d0a.
129. Vincent, J. 2018. IBM secretly used New York's CCTV cameras to train its surveillance software, *The Verge*, September 6, 2018. Available at www.theverge. com/2018/9/6/17826446/ibm-video-surveillance-nypd-cctv-cameras-search-skin-tone.
130. Currier, C. 2016. A walking tour of New York's massive surveillance network. *The Intercept*, September 24, 2016. Available at https://theintercept.com/2016/09/24/a-walking-tour-of-new-yorks-massive-surveillance-network/.
131. Chen, S. 2019. The Chinese technology helping New York police keep a closer eye on the United States' biggest city. *South China Morning Post*, January 11, 2019. Available at www.scmp.com/news/china/science/article/2181749/ chinese-technology-helping-new-york-police-keep-closer-eye-united.
132. Chinoy, S. 2019. We Built an 'unbelievable' (but legal) facial recognition machine. *New York Times*, April 16, 2019. Available at www.nytimes.com/ interactive/2019/04/16/opinion/facial-recognition-new-york-city.html.
133. Renzulli, K. A. 2019. Here are the 15 jobs disappearing the fastest in the US. *CNBC*, April 28, 2019. Available at www.cnbc.com/2019/04/26/the-15-us-jobs-disappearing-the-fastest.html.
134. Huckle, S., Bhattacharya, R., et al. 2016. Internet of Things, blockchain and shared economy applications. *Procedia Computer Science*, 98, 461–466.
135. Zhang, Y. and J. Wen. 2017. The IoT electric business model: Using blockchain technology for the Internet of Things. *Peer-to-Peer Networking and Applications*, 10(4). doi:10.1007/s12083-016-0456-1.
136. Detwiler, B., 2019. Autonomous cars as a service: How Panasonic is preparing for a future where most drivers don't own cars. *Tech Republic*, January 28, 2019. Available at www.techrepublic.com.
137. Kshetri, N. 2017. Can blockchains strengthen the Internet of Things? *IT Professional*, 19(4), 68–72.
138. Li, C. and L.-J. Zhang. 2017. A blockchain based new secure multi-layer network model for Internet of Things. *Internet of Things (ICIOT), 2017 IEEE International Congress on*, IEEE, pp. 33–41.
139. Kravitz, D. W. and J. Cooper. 2017. Securing user identity and transactions symbiotically: IoT meets blockchain. *Global Internet of Things Summit (GIoTS)*, IEEE, pp. 1–6.
140. Huh, S., Cho, S. and S. Kim. 2017. Managing IoT devices using blockchain platform. *Advanced Communication Technology (ICACT), 2017 19th International Conference on*, February 19–22, 2017, Bongpyeong, South Korea.
141. Samaniego, M. and R. Deters. 2016. Using blockchain to push software-defined IoT components onto edge hosts. *BDAW '16 Proceedings of the International Conference on Big Data and Advanced Wireless Technologies*. Article No. 58, Blagoevgrad, Bulgaria—November 10–11, 2016. ACM, New York. doi:10.1145/3010089.3016027.

142. Lee, B. and J.-H. Lee. 2017. Blockchain-based secure firmware update for embedded devices in an Internet of Things environment. *The Journal of Supercomputing*, 73, 1152. doi:10.1007/s11227-016-1870-0.
143. Jones, D. 2019. AT&T and IBM: Tackling A $1 trillion opportunity. *Seeking Alpha*, July 17, 2019. Available at https://seekingalpha.com/article/4275521-t-ibm-tackling-1-trillion-opportunity.
144. Jeyanthi, N. and R. Thandeeswaran. 2017. *Security Breaches and Threat Prevention in the Internet of Things*, IGI Global, Hershey, PA.

Chapter 3

A Blockchain-enabled edge supported e-challan mechanism for content-centric Internet of Vehicles

Ali Nawaz Abbasi, Tashjia Anfal,
and Muhammad Maaz Rehan

CONTENTS

3.1 INTRODUCTION

Vehicular ad hoc networks (VANETs) are getting the attention of researchers to contribute to building a smart transportation system, improvising road safeties to ensure safe driving, disbursement of traffic conditions to ensure hassle-free driving towards the destination, and most importantly, encouraging towards autonomous vehicles [1]. Several protocols, solutions, and applications, like Dedicated Short Range Communication (DSRC) [2] and wireless access in vehicular environment (WAVE) [3], were proposed to achieve these milestones. The Internet of Things (IoT) is reshaping homes to smart homes, health services to smart health, and VANETs to the Internet of Vehicles (IoV). The evolution from VANETs to IoV is the result of connecting vehicles with the Internet, which changed the paradigm, to not just limit the communication with infrastructure for road safety but also to ensure the provisioning of infotainment, payment services, and other relevant applications [4].

3.1.1 VANETs versus IoV

Let's discuss how IoT has reshaped VANET into IoV and what are the additional IoV features. In [5,6], comprehensive comparative details are presented which portray the limitations of VANETs. In VANETs, communications is only in the form of vehicle to vehicle (V2V), vehicle to road (V2R) side units, and vehicle to infrastructure (V2I), while IoV expanded the vehicle communication to sensors (V2S), human/personal devices (V2H/V2P), and with everything (V2X). By studying comparative studies, it can be observed that there is an addition of sensors which cooperate in the transmission of vehicle sensory data to monitor health and violations, personal devices to communicate traffic and transportation information, and involving cloud services to manage the heterogeneity of application and services. Because of limited communication architecture, VANETs are small networks comprising vehicles and roadside units (RSUs). Although IoV involves devices which are capable of communication, it has made this architecture scalable and heterogeneous. VANETs serve a limited number of services in small networks, while IoV is capable of hosting a big number of applications and services among the devices on a large scale. These all are possible because of communication technologies which served as the backbone. If we talk about VANETs, along with DSRC and WAVE, continuous air interface for long and medium range (CALM) is also in the count of VANETs network technologies. IoV builds its communication on Bluetooth, ZigBee, 4G/LTE, and WiMax which are accompanied by VANETs network technologies.

3.1.2 Content-centric IoV

The aforementioned technology advancements are the fair resultants of 'Internet', which quickly leads towards the client-server connectivity based on

Internet protocol (IP) addresses, and the transfer of data packets is dependent on the stable communication link. Over time, increasing the growth rate of vehicles will also impact data traffic, the majority of which currently lies in aforesaid IP-based connectivity. There is a prediction by Cisco visual networking index: The increase in IP traffic will reach the figure of 396 Exabytes by 2022 [7], which will introduce high delay and congestion as a result of host-centric networking. Therefore, it is a future requirement to ultimately switch from host-centric communication. Existing IP-based connection needs to first establish a stable connection between nodes for the successful transfer of data, which could be affected by the fast mobility of vehicles. Because of this route establishment and maintenance, developing a reliable vehicular network is a difficult and complex task [8]. To tackle all of these discrepancies, content-centric networking (CCN) is emerging as a promising future Internet architecture used by several researchers to minimise the delay in content provisioning and effectively dealing with dynamic link connection requirements because of the mobility of vehicles [9–11]. Because of the distributed nature of CCN, the content could be cached at any node in the network, which provides an ultimate alternative to the problem of flooding interest towards only content holders in an IP-based Internet architecture. CCN uses an interest-data topology among provider, source, and router [12], in which consumer demands content by publishing an interest packet regardless of the location of the content, and content could be cached anywhere in the network.

Figure 3.1 depicts the content-centric Internet of Vehicles (CC-IoV) operation. The consumer floats an interest packet with a content name on a receiving interest packet if nodes will search content in their content store (CS). This does away with the need of establishing a connection, but the question arises whether the content provided is trustworthy or provided by a malicious node to mislead the consumer, which shows that in this paradigm content correctness and originality is an utmost requirement and necessitates deploying all these with some auditable and trustworthy content and content provider ledger (i.e., Blockchain).

3.1.3 Blockchain with edge computing

Blockchain [13] was first introduced by S. Nakamoto for cryptocurrency Bitcoin. An immutable and auditable distributed append-only ledger, Blockchain proves itself a trustworthy database. Complex calculations, use of exhaustive consumption of hardware resources, and distributed copies of data have played an important role in the quick rise of Blockchain. Because of these properties, Blockchain has attracted researchers in relevant fields for a tamper-proof and trusted data organisation. A number of verified transactions are accumulated to form a block, and afterward, a HASH is calculated for the block. Each block is linked with the previous block's HASH, which forms Blockchain.

Figure 3.2 illustrates a working overview of transaction verification, block creation, and block addition in Blockchain. A block is not just directly added

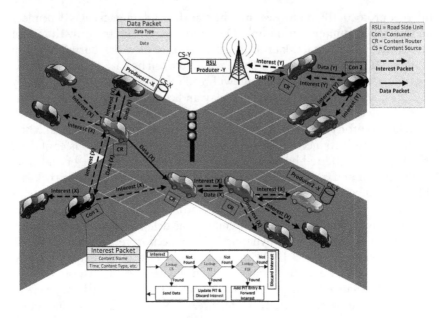

Figure 3.1 Interest-data operation in content-centric Internet of Vehicles.

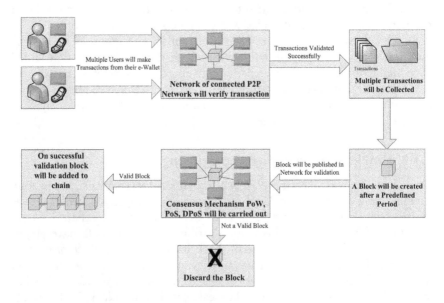

Figure 3.2 Operational overview of Blockchain.

into the Blockchain, it is first validated by the 'group of nodes-Blockchain Network', miners, following a 'Consensus Mechanism'. A Blockchain network could be permissioned (private) or permission-less (public); a permissioned Blockchain is deployed over a group of authorised nodes which participate in block mining and validation in the Blockchain network, while on the other hand, any node can participate in permission-less Blockchain [14].

Blockchain is a decentralised ledger and needs exhaustive computations on every individual Blockchain node. In CC-IoVs, RSUs can be featured with edge computation, which will serve for service requests and data provisioning among service and application providers and vehicles. Deploying a distributed Blockchain on RSUs enables frequent calculations and data storage. In Blockchain-enabled edge computing, RSUs are the edge computing nodes that receive data or service requests and act on it respectively. Another activity is to gather transactions and ensemble them in the form of a block by calculating the HASHes, which on successful validation from the Blockchain nodes merged in the Blockchain. In Figure 3.3, Blockchain-enabled CC-IoV is depicted, in which RSUs are equipped with the feature of edge computing and ending up as in the shape of Blockchain-enabled edge node (BEN).

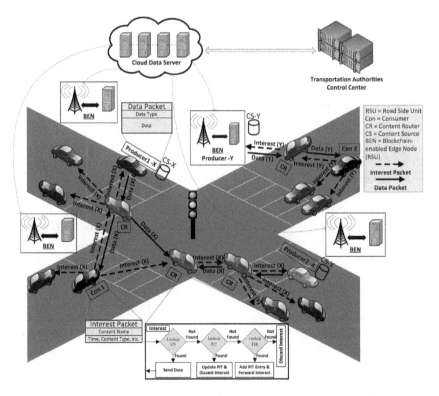

Figure 3.3 Operational overview of content-centric Internet of Vehicles (CC-IoV) with Blockchain-enabled edge computing nodes.

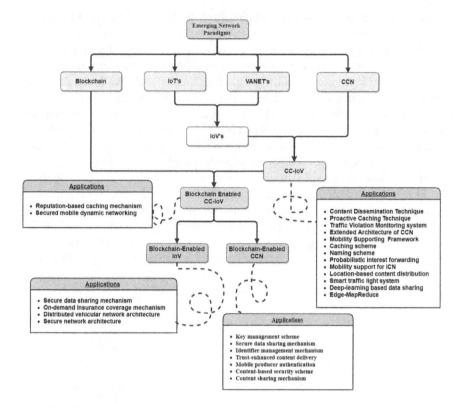

Figure 3.4 Emerging network paradigms.

As data transmission is open to access for all of the receiving entities because immutability and audit ability, Blockchain is a proven potential technology to be used among vehicular network entities to ensure that no malicious data activity will take place, and if it happens, then it would be apprehended easily because of Blockchain traits. Figure 3.4 enlightens the emerging network technologies with their applications. Later in this chapter, we will discuss the use of both CCN and Blockchain in IoV in referred research studies, then on behalf of literature study, research motivation will be discussed, and then we will present our work to overcome the missing issues not discussed in the existing works. In the results and discussion section, we will validate our solution on behalf of the results collected through running simulations with a set of parameters.

3.2 LITERATURE SURVEY

In this section, existing work in the field is discussed and has been divided into two parts. The first part is related to CCN-based IoV. The second part is related to Blockchain-enabled technologies which are further divided into three parts: The first subsection discusses Blockchain-enabled CCN, the second subsection is related to Blockchain-enabled IoV, and the last subsection focuses on Blockchain-enabled CC-IoV.

3.2.1 Content-centric Internet of Vehicles (CC-IoV)

In [15], the authors have proposed a monitoring system for traffic violations for VANETs based on CCN. Interest and data transfer takes place among cop vehicles (CV) and every ordinary vehicle (OV). OVs maintain a record of violation history based on sensor data from on-board units (OBU). Whenever an OV passes by a CV, the data packet is offloaded to CV against the interest packet of CV. This makes it possible to monitor and control violations in unpatrolled and uncovered areas. But this leaves open the issues of trust management among CV and OV, record management of violations for cross-validations, and repeat ticket issuance from another CV of the same violation.

In [16], the authors have proposed a proactive caching technique with the vehicular CCN concept to minimise delay and to deal with dynamic connection requirements resulting from mobility. A car sends its interest to data store (DS) through RSUs; DS divides the data requested into chunks and finds out RSUs and their distances to consumers. After these calculations, DS places the chunks onto forthcoming RSUs in the direction of the car. The technique focuses on personal data and V2I communication remaining in centralised fashion, and the mobility of DS is also not taken into account which is needed for smart vehicular systems.

In [17], the authors have proposed content dissemination techniques to control broadcast storm and flooding. This technique uses the geo-location to suppress the forwarding of a redundant interest packet. On receiving an interest packet, the vehicle calculates the distance from the requester, and the vehicle which is farthest will set the lowest deferral timer and will forward the interest first. On reception of interest from forwarder node, nodes will calculate the direction of the forwarder, moving towards the requester and then stop the timer to forward interest or otherwise forward the interest. This will check other nodes to forward the interest packet. However, no caching strategy is discussed to minimise delays and hops.

In [18], the authors have proposed a caching scheme for content-centric vehicular networks. Vehicles form a group based on most spent time in a region, and the group is named 'hot region'. Vehicles share their contents containing their coordinates and timestamp to RSUs. Vehicles in a specific hot region cache the contents of neighbouring vehicles and provide content from their content stores on demand. A cache replacement policy is also followed which depends on the popularity of the content; low popularity content is replaced with a new one or popularity is updated. This ensures minimisation of delay and minimises hops from the content provider, but a trust level of provider is missing and cache pollution can be accomplished through a Sybil attack.

In [19], the authors have proposed a naming scheme using geolocation for content-centric vehicular networks to minimise broadcast storms in networks. The naming scheme for the interest packet is designed to notify those nodes which are moving towards, or in the vicinity of, the point of interest. Coordinates for the point of interest are integrated into the content name of the interest packet, and only the aforementioned entities will forward the interest packet. This helps to avoid the broadcast storm but leaves delay as an open issue because data could be several hops away from the consumer.

In [20], the authors have proposed an extended architecture of CCN to cope up with the dynamic connection requirements of vehicular networks. An additional data store named a 'neighbour content store' is used to cache a neighbour's critical messages in a few hops of neighbouring vehicles of the producer. This caching strategy to store content is done by using the 'time to live; field. This extension in CCN introduced data and content existence on a few hops of the producer which minimised delay. But no security mechanism for content is opted to avoid interference of malicious nodes.

In [21], the authors have proposed a mobility supporting framework for vehicular named-data networks (VNDN). This technique focuses on V2V communication, which introduces the mobility of content sources. Forwarding redundant interest and data packets are restrained by involving each node to calculate the distance from the sender and setting a timer, and on the reception of the forwarded packet, each receiving node checks other forwarding nodes. The design includes anchor zones (AZ), any node in AZ replicates the contents in CS, and on joining a new AZ, nodes replace old content with new content which ensures content on fewer hops and minimises delays. But trust is an open issue among nodes.

In [22], the authors have proposed a probabilistic technique for interest forwarding in CCVN that mitigates broadcast storms. In the proposed technique, each vehicle dynamically calculates the probability that is based on their current neighbour's density. Vehicles do not inform about their geographical location to their neighbours or content producers, and vehicles do not have any precedent information about network topology. By modifying existing PIT of traditional CCN, the local information can obtain and guess the count of current neighbours in the surroundings by using local density approximation technique. Moreover, an interest rebroadcasting mechanism is given in which the time-based scheme is used that prioritises the potential forwarder on the bases of distance from the sender.

In [23], the authors provided mobility aid services via fog computing with information-centric Internet of Vehicles (IC-IoV). The provided support uses the data features categorised furthermore into user shareable and non-shareable data. These defined characteristics of data along with computation and geographic information of fog computing are used to design a scheme for communication and data sharing while making use of provided IoV applications such as alarms in danger and updating the information about traffic and communication in V2V scenarios. The proposed scheme enables distinguished mobility support for moving vehicles for consumer and producer mobility, respectively, to care to take two completely different associated mobility problems. Moreover, shortcomings of continual upgradation of FIB are discussed, which causes packet loss and higher delay in terms of requesting and retrieving desired data. However, the proposed mechanism faces a trade-off in terms of higher performance and cache or storage deployment cost of fog nodes.

In [24], the authors have proposed the content distribution mechanism based on the location named LoICen that mitigates the broadcast storm. The proposed LoICen mechanism holds the location and mobility information collected opportunistically of the vehicles that have content in their cache for better content delivery. The obtained geographical information can be used at any time for better content search by selecting the most competent neighbouring node or vehicle on every individual hop. Information based on location can be used for routing interest packets towards the area where content is cached and also for content discovery. Moreover, location-based forwarding of the interest packet for the requested content is proposed. In which any intermediate node or vehicle that has requested content in their cache can stop the forwarding of interest packet by replying with the requested content chunks to the requestor. Furthermore, the proposed scheme minimises the delay of retrieving the content and count of interest packet transmissions for a content search.

In [25], the authors proposed a smart traffic lights system for VNDN, in which packet propagation is based on geolocation mechanism. The proposed mechanism resolves traffic congestion problems along with minimising the vehicles waiting on the road in a decentralised manner. The existing traditional system of traffic lights is replaced with virtual traffic lights (VTLs). In which RSUs behave like intersection controllers that control the intersections instead of typical road signals. RSUs only entertain the vehicles found in their defined zone by accumulating vehicle order based on the priority mechanism that has reached or will reach the intersection. After that, the RSU forwards an immediate message to each vehicle in which it informs the vehicles to wait for some time (red signal) or move (green signal) from the intersection. Furthermore, the RSUs share information with neighbouring RSUs about the vehicles that could arrive at their vicinity; this will help to make decisions on time about managing road traffic. The proposed system is more ideal in autonomous vehicles as they only get messages from RSUs rather than processing images.

In [26], an edge-MapReduce method to perform route planning in autonomous vehicles is introduced. The architecture analyses the big data acquired in the IC-IoV. In the proposed model, the route sequence makes use of intelligent transportation system (ITS) node, edge node, mapper, and reducer. The edge node repeatedly requests the real-time traffic information in the interest packet and is answered by the intelligent node in the network. The data packet is further directed towards the mapper; in response to the topology, the weight map interest packet is forwarded towards the edge node. The reducer also seeks the topology weight map data from the mapper which is addressed while returning the data packet in response. The edge node then inquires for real-time user demand requests in the interest packet and collects the data from the intelligent node. The reducer further collects the user demand data packet after sending the interest packet to the edge node. Finally, the reducer provides the route assignment information to the intelligent node in the ITS through the edge node.

In [27], the data dissemination scheme based on a deep learning mechanism is introduced for CC-IoVs. The proposed mechanism comprises three stages that consider vehicle mobility along with the types of shareable data. The first stage specifies the criteria of participant vehicles that can contribute to data distribution based on the energy estimation mechanism. In the next stage, the Weiner process model is used to discover stable and authentic connections by computing vehicles connecting probability. At the final stage, the evaluation design based on a convolutional neural network (CNN) is defined in which social relationships between vehicle pairs are rated. Furthermore, the most suitable vehicle pair is distinguished using CNN, which ensures data sharing with minimum delay along with a more available data rate.

3.2.2 Blockchain-enabled CC-IoV

This part is divided into three subsections as explained.

3.2.2.1 Blockchain-enabled IoV

In [28], the authors have proposed a semi-automated on-demand insurance coverage mechanism using Blockchain as a distributed ledger for smart contracts of insurance policies among insurance companies and consumers. A user can activate and deactivate or modify insurance coverage among theft, accidental, passenger, or weather by just using a mobile app, and all the status related to each coverage type is transmitted to the server through an electronic device which is responsible for gathering sensor data. To modify insurance or cancel insurance claims, sensor data plays a vital role. All these modifications are done according to the smart contract, and these changes are stored on the Blockchain which provides immutable and auditable records.

In [29], the authors have proposed a solution to tackle secure data sharing in vehicular communications. A trust level is developed by integrating

Blockchain to keep a record of message transactions of traffic conditions, penalty or rewards. Traffic management authority (TMA) is a key responsible role in maintaining privacy about the identities of vehicles. The trust level is built by a signature threshold; a message is trusted if the number of vehicles crossing the threshold has signed the message. Incentivising is also introduced to push contributors to remain honest. Broadcasting messages periodically with no forwarding strategy introduces network congestion and delay.

In [30], the authors have proposed Blockchain-based vehicular network architecture to acquire the benefits of decentralised networks. Enabling Blockchain in the network enables trusted data sharing among users. Because of a variety of data in applications, Blockchain is classified into different blockchains. Vehicles take part with other network devices to generate blocks with their key pair embedding in the transaction. This introduces frequent block generation which introduces data traffic flooding in the network; also the architecture remains unanswered about 'who will and how the transactions and blocks will be validated'.

In [31], the authors proposed Blockchain-based a secure network architecture for IoVs in which vehicles can communicate and share network resources along with information in a secure manner. The presented algorithm permits storing of multiple data copies on a distributed storage system that make the data more secure than conventional centralised mechanism. The proposed scheme comprises inception and encryption along with data migration. Moreover, block building construction with decryption and data uploading is included in the algorithm. In inception, the IoV network generates a unique ID for each user for secure communication. After that encryption process is enabled by the IoV network when a user wants to transfer the data. Then the block is constructed by the network that has a hash value associated with a primary key and broadcasts this block after the verification and validation of hash value. The new block is added by all the nodes of the IoV network to their Blockchain. At last, the user follows a decryption process based on the public Blockchain ledger approach and may retrieve its desired data. However, the proposed framework increases data retrieving delay, but it provides security along with reliability and trust among connected vehicles and helps them in decision making by exploiting IoV applications through a distributed control system.

3.2.2.2 Blockchain-enabled CCN

In [32], the authors have proposed a key management scheme to authenticate roles in named data networks (NDN) by storing keys on Blockchain. This ensured authenticity and built a trust level on producers of content. Every node, including devices and applications, registers themselves with their name prefixes and public key in the form of transaction which is validated by a group of private Blockchain nodes, and afterward, a transaction is added to the Blockchain. The user is authenticated by its block height, transaction hash, and status of valid or invalid. Always establishing a link with site administrator every time for authentication is time-consuming and introduces a delay.

In [33], the authors have proposed a secure data-sharing mechanism in CCN, which also preserves privacy. Blockchain is integrated to build a trust level and auditable ledger. Every user in the network is registered against its unique ID and key pair. Nodes place the pointer of contents in transactions, which are placed on Blockchain after validation from a set of miner nodes selected through a voting event among users of common social interest groups. The miner who validates the block gets the content from providers as a reward. A consumer can get the content by 'passing' access control policy set by the provider which introduces excess network use and delay.

In [34], the authors have proposed an identifier management mechanism that is secured by the use of Blockchain. Blockchain is used in a hybrid way by using both public and private Blockchain. A new node can register itself on a private Blockchain with its key pair and can play a role in the verification of transactions and creation or validation of blocks. Content in the NDN environment can be delivered to the consumer without exposing the ID of a producer, which saves the entity from the denial-of-service (DoS) attack. But it is not an option to check a malicious node or a group of malicious nodes to join the network because any node can register itself with a key pair.

In [35], the authors have proposed Blockchain-based content delivery with enhancing trust for information-centric networks (ICNs). The content provider registers its content objects in the form of transactions on a logical Blockchain host (LBH) through name resolution executors (NRE) and gets a payload registered on LBH against a transaction ID. The consumer requests the content object by a discover message which travels through the NRE and then a parent NRE until it reaches NRE where the transaction is registered. A reverse path through content routers is established, and content is sent to a consumer. A countermeasure to ensure either transaction is part of the Blockchain is taken, which ensures trust because of immutable and auditable distributed ledger (i.e., Blockchain). The only route to the content provider is stored, which introduces a delay in content delivery.

In [36], the Blockchain-based distributed mobile producer authentication mechanism named 'BlockAuth' is introduced for ICN. The proposed scheme addresses ICN mobility management problems that lead to security challenges and provides secured and high-speed mobile user authentication. BlockAuth exploits a weighted clustering algorithm (WCA) in which the cluster heads of the mobile nodes are selected on the bases of mobility, transmission, and battery power. The scheme categorises networks in the form of the clusters (such as micro-cells) and core routers. Each cluster has a cluster head (CH) and the number of access gateways, for example, Wi-Fi access points or base stations of ICN. The CH manages all the transactions happen between the cluster members and mobile producers connected with their particular cluster. While in the network, core routers that manage the Blockchain are known as 'global Blockchain administrators' (GBAs). These routers cache all the transactions done by the clusters for intercluster handover. Also, these routers collaboratively supervise the incoming and outgoing packet exchange of the participant routers including the CH while handling the mobile producer's transaction process that is done by different clusters.

In [37], the authors proposed a content-based security scheme called a 'decentralised public key infrastructure' (DPKI) that uses the Blockchain mechanism for CCN. The proposed scheme overcomes the public key infrastructure (PKI) existing model limitations by eliminating the reliance on a centralised authority (CA). DPKI comprises a two-step registration and verification. In registration, the owner of the content registers its identity with the corresponding public key and then sends a message to the Blockchain along with its encrypted private key. The Blockchain miners verify the identity, the nodes validate the transaction, and then this identity is added into the next block. In verification, when the content owner registers itself, its identity and the public key becomes public. So, the user can verify its identity and public key against each transaction by traversing through Blockchain lookup. Also, transactions are digitally signed and registered after verification, which makes tampering difficult in a public key.

In [38], the authors proposed a content-sharing mechanism by enabling Blockchain in VNDN. In the proposed scheme, data sharing is based on Blockchain that comprises a double layer. Nodes on the first layer announce their demands on the NDN paradigm and requests services. While on the other layer, nodes submit their requests and forward it to the closest RSU for more data. Furthermore, the similarity game model is constructed for one to many along with a reasonable Reputation Evaluation Scheme (RES) model that prevents the dissemination of false data from evil nodes. However, the proposed technique promotes the secure dissemination of data between nodes, but it increases the computational overhead.

3.2.2.3 Blockchain-enabled CC-IoV

In [39], the authors have proposed a caching scheme based on the reputation of the content provider in VANETs. The caching scheme depends on the reputation of the content provider; already existing low reputation value content will be replaced with the content of a high reputation value content provider. The reputation value of network nodes is maintained over the Blockchain network which uses the proof-of-work (PoW) consensus mechanism. A consumer can increment or decrement the reputation value of producer on the reception of valid or invalid content. No mechanism is opted for restricting the malicious nodes to be the part of the network, which makes the network vulnerable for reputation pollution.

In [40], the authors have proposed to integrate Blockchain with vehicular CCN to get decentralisation, secured, and mobile dynamic connections.

Contents like cooperative awareness messages (CAM) from vehicles should be covered up and listed in transactions, which will be the part of the block. Afterward a block will be added to Blockchain to get a distributed and trusted ledger. Adding all of the CAM as transactions are overhead of transmission, keeps miners busy with Blockchain management, and no consensus mechanism has been explicitly opted.

3.3 RESEARCH MOTIVATION AND PROBLEM STATEMENT

The literature survey raises some major queries:

- How to record traffic offenses in the absence of traffic wardens and cameras at any time of the day?
- How to develop such a distributed trusted environment for CC-IoV that holds offender trust as well?
- How to report the offensive events and prove them?

Based on the literature review reported in the literature survey, conclusions have been drawn (Table A1.1).

An auditable and trustworthy ledger of content sharing based on Blockchain hardly exists for the CC-IoV.

CC-IoV [16–21] work does not provide an auditable ledger of contents. The work in [15] has proposed a database of offenses along with patrolling traffic police vehicles. The Blockchain-based schemes [32–34] do not record the content shared among entities and that wastes the effective use of Blockchain in the scenario of a content-centric environment. The work in [35] proposed to record content object pointer in Blockchain which lacks in the trust issues because of the loss of data on producer end.

The proposed mechanisms in [28–30] are IP-based which need a stable connection to be established first. Although the proposed solutions in [39,40] have missed the essence of Blockchain in CC-IoV, provisioning of an auditable ledger of content shared to push entities towards honest participation in the network for content sharing.

The proposed edge-computing-based cooperative transaction validation for offensive events in Blockchain-enabled CC-IoV aims to achieve the following objectives:

- To develop a cooperative scheme that detects the offense by involving the BEN and vehicles and then raise a report against the offender.
- To develop a trusted transaction validation scheme in a distributed CC-IoV environment that performs validation with minimum delay and least control overhead.

3.4 PROPOSED BLOCKCHAIN-BASED EDGE MECHANISM FOR CC-IoV

This section presents the proposed scheme for the validation of events to transaction formation. First, it is worth mentioning and inculcating that data transfer among entities takes place with the help of packets transmission; each packet contains specific information from the sender, while the receiver makes some decision and steps over packet type. Packet formation and contents will be discussed in this section, which will allow receivers to make some decisions in completing the process of an event occurring to transaction validation for Blockchain. Then, we will provide an overview of the scheme in steps.

3.4.1 Generating packets

In CCN, content sharing takes place between producers and consumers using interest or data packets transfer. In our scheme, we have targeted the speed offense as the offending event and a key stakeholders' producer is 'offender node' because it generates the information of speed offense on collection of sensory data; consumers are the 'dealing authorities or TMA' getting information prompted by the edge computing node and consuming the information to record offensive events and generate e-tickets for offenders. To achieve a fair and trusted working scheme, the details of packets shared and steps of data sharing follow.

3.4.1.1 Speeding vehicle info delivery

We assume that vehicles are equipped with speed sensors attached to their OBU. If a vehicle crosses over the speeding thresholds, OBU will start advertising its vehicle information packet until the offender's speed is not inside the limits specified by the authorities; initially the packet contains the vehicle identification and speed. As the packet broadcast will only initiate after the speed violation, there will be no overhead on the network unless an offender commits the violation. One-hop neighbours will be the receiver of this packet. Receiving vehicles will start counting the vehicle information packet. If the received count is within specified limits, it will take no action considering an overtake activity or other emergency conditions; it will, however, generate and send offense reporting packet. In Figure 3.5, the packet formation and contents used in the delivery of offender vehicle information which helps to calculate and allocate bandwidth for data transmission is shown.

3.4.1.2 Reporting an offense

As receiving of the VehInfoPkt is counted, if received count crosses the threshold, it ensures the vehicle is continuously violating the speed regulation

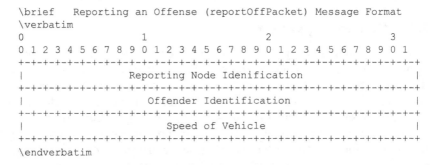

```
\brief   Vehicle Information (VehInfoPKT) Message Format
\verbatim
0                   1                   2                   3
0 1 2 3 4 5 6 7 8 9 0 1 2 3 4 5 6 7 8 9 0 1 2 3 4 5 6 7 8 9 0 1
+-+-+-+-+-+-+-+-+-+-+-+-+-+-+-+-+-+-+-+-+-+-+-+-+-+-+-+-+-+-+-+-+
|                    Vehicle Idenification                     |
+-+-+-+-+-+-+-+-+-+-+-+-+-+-+-+-+-+-+-+-+-+-+-+-+-+-+-+-+-+-+-+-+
|                      Speed of Vehicle                        |
+-+-+-+-+-+-+-+-+-+-+-+-+-+-+-+-+-+-+-+-+-+-+-+-+-+-+-+-+-+-+-+-+
\endverbatim
```

Figure 3.5 Packet formation of VehInfoPKT.

```
\brief   Reporting an Offense (reportOffPacket) Message Format
\verbatim
0                   1                   2                   3
0 1 2 3 4 5 6 7 8 9 0 1 2 3 4 5 6 7 8 9 0 1 2 3 4 5 6 7 8 9 0 1
+-+-+-+-+-+-+-+-+-+-+-+-+-+-+-+-+-+-+-+-+-+-+-+-+-+-+-+-+-+-+-+-+
|                  Reporting Node Idenification                |
+-+-+-+-+-+-+-+-+-+-+-+-+-+-+-+-+-+-+-+-+-+-+-+-+-+-+-+-+-+-+-+-+
|                    Offender Identification                   |
+-+-+-+-+-+-+-+-+-+-+-+-+-+-+-+-+-+-+-+-+-+-+-+-+-+-+-+-+-+-+-+-+
|                      Speed of Vehicle                        |
+-+-+-+-+-+-+-+-+-+-+-+-+-+-+-+-+-+-+-+-+-+-+-+-+-+-+-+-+-+-+-+-+
\endverbatim
```

Figure 3.6 Packet formation of reportOffPacket.

set by authorities or else some other activity like overtaking other vehicles is done by the packet originator. After the confirmation of the speed offense over packet count, the receiving neighbour will generate a report offense packet for BEN in its transmission range as shown in Figure 3.6. The same packet contains the reporting vehicle identification, offender's identification, and speed at the time of the offense. Packet formation and contents with their data length are depicted in the following figure.

3.4.1.3 Validation of offense

As the report offense packet is only for BEN, only BEN will act on receiving the report offense packet, while the rest of the receiving nodes will discard the packet. BEN will maintain a local database to save the partial information of reporting vehicle identification, offender's ID. and speed of offender, which will be the part of the transaction after a validation process. Transaction validation will be performed by asking the neighbouring nodes of BEN, which have recorded the offense of the reported offender in their local database. BEN will generate and advertise an interest packet validating the offense. Figure 3.7 depicts the packet formation and content data length for offense validation.

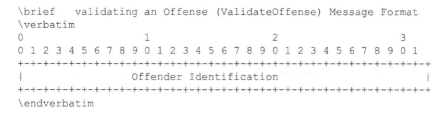

```
\brief    validating an Offense (ValidateOffense) Message Format
\verbatim
0                   1                   2                   3
0 1 2 3 4 5 6 7 8 9 0 1 2 3 4 5 6 7 8 9 0 1 2 3 4 5 6 7 8 9 0 1
+-+-+-+-+-+-+-+-+-+-+-+-+-+-+-+-+-+-+-+-+-+-+-+-+-+-+-+-+-+-+-+-+
|                    Offender Identification                   |
+-+-+-+-+-+-+-+-+-+-+-+-+-+-+-+-+-+-+-+-+-+-+-+-+-+-+-+-+-+-+-+-+
\endverbatim
```

Figure 3.7 Packet formation of ValidateOffense.

3.4.1.4 *Reply of offense validation*

On receiving an offense validation packet, vehicles will look up in their local data store for the existence of offender details. On successful retrieval of a matching record, vehicles will generate a reply packet for the validation of the offense confirming that the offender has violated the speeding regulation set by authorities and will advertise the packet targeting BEN. On receiving the reply packet of offense validation, officer node will merge the confirming vehicle's information in matching the offender partially filled transaction. The packet formation of the reply validation packet is shown in Figure 3.8.

Finally, when m-numbers of transactions are collected by edge computing node, it will generate a block and will broadcast the block in the network for validation, which after successful validation is merged in the Blockchain.

3.4.2 System workflow of scheme

A summarised workflow of the scheme keeping the speeding offense in view with steps is as follows and is also depicted in the Figure 3.9:

Step-01: Vehicles will continuously monitor their speed sensor data. On violation of speeding limits, the vehicle will periodically advertise its identification and speed data by VehInfoPkt.

Step-02: On receiving VehInfoPkt, vehicles will maintain a record with the count of the packets received. If the packet received count reaches

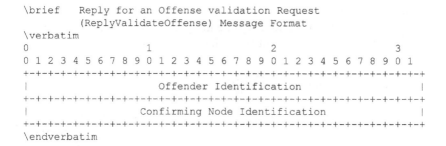

```
\brief    Reply for an Offense validation Request
          (ReplyValidateOffense) Message Format
\verbatim
0                   1                   2                   3
0 1 2 3 4 5 6 7 8 9 0 1 2 3 4 5 6 7 8 9 0 1 2 3 4 5 6 7 8 9 0 1
+-+-+-+-+-+-+-+-+-+-+-+-+-+-+-+-+-+-+-+-+-+-+-+-+-+-+-+-+-+-+-+-+
|                    Offender Identification                   |
+-+-+-+-+-+-+-+-+-+-+-+-+-+-+-+-+-+-+-+-+-+-+-+-+-+-+-+-+-+-+-+-+
|                 Confirming Node Identification               |
+-+-+-+-+-+-+-+-+-+-+-+-+-+-+-+-+-+-+-+-+-+-+-+-+-+-+-+-+-+-+-+-+
\endverbatim
```

Figure 3.8 Packet formation of ReplyValidateOffense.

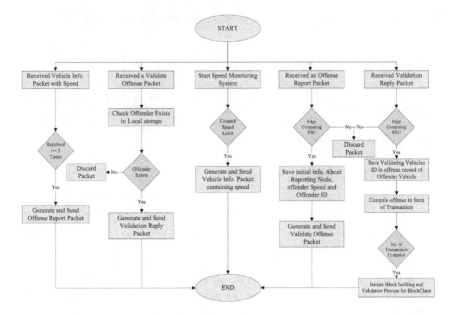

Figure 3.9 Working schema of the proposed system.

the threshold, the offense will be reported to BEN with generating and broadcasting a reportOffPacket which contains the offender identification, violation speed, and reporter's identification for future references.

Step-03: BEN will compile a new transaction with partial information extracted from the offense report packet and will validate the offense report by generating and sending an offense validation packet to validate the transaction.

Step-04: Validation request and interest receiving vehicles will generate and send a reply for offense validation by searching the existence of the offender details in their data table.

Step-05: BEN will add the confirming vehicle ID in the transaction for matching the offender ID.

Step-06: Finally, after m-number transactions, edge computing node will generate a block and will be merged in the Blockchain after successful validation consent from the Blockchain network.

3.5 RESULTS AND DISCUSSIONS

Here we will briefly discuss details of the simulation environment and results for our scheme. First, we will discuss the details of our simulation environment and parameters used to verify the correctness of our work. Then we will compare our proposed work with technique in [15] to verify that our proposed scheme is better in performance and trust issues among stakeholders for CC-IoV than the referred technique.

3.5.1 Simulation environment

We have evaluated our proposed work using Network Simulator 3 (ns-3) [41] over Ubuntu 16.04 LTS [42]. A real VANETs mobility pattern was generated by using simulation for urban mobility (SUMO) [43]. For this work, the 26 No. area of Islamabad was selected for generating a VANET mobility pattern. In Figure 3.10 is the screenshot of the 26 No., Islamabad taken from Google Maps. The next step for generating the pattern mobility through SUMO was to export a road map from OpenStreetMaps (OSM) in the '.osm' format.

Then using SUMO '.tcl' scripts containing the mobility patterns of 30, 60, and 90 Nodes were generated, which were further used in NS-3 to simulate the real-time environment for VANETS. Following is the generated SUMO map for exported '.osm' file of 26 No. area of Islamabad. This generated mobility pattern was then fed to our compiled simulation to evaluate our proposed work. We have used the ad hoc on-demand distance vector (AODV) routing protocol for simulation purposes while the rest of the simulation parameters are listed in Table A1.2.

3.5.2 Results and analysis

In this section, we will provide a comparative analysis of our proposed scheme with Traffic Violation Ticketing (TVT) in graphical form.

3.5.2.1 Time from offense to Blockchain transaction

Time from offense to Blockchain transaction is being defined as a delay of recording an offense from its occurrence to data storage. We have calculated this as the average time of five offenses on the setting of 30, 60, and 90 vehicles. As the number of vehicles increases, participation for reporting and validating an offense also increases, which impacts the increasing of delay.

Figure 3.11 demonstrates the comparative delay of our proposed work and TVT. TVT depends on the nearest CV to share its offense, while in our proposed work vehicles work cooperatively and do not need to wait for a special entity to offload its offense record, which has minimised the delay. The graph depicts that our proposed work has the least delay than TVT even increasing vehicles.

3.5.2.2 Packets overhead

Packets overhead are being defined as the additional packets created and used to identify an offense, report, validate, and then finally collect validation replies from different nodes that have recorded an offense of the same

Figure 3.10 Google map of 26 No. area located in Islamabad, Pakistan.

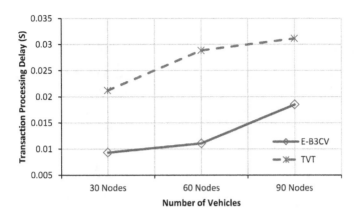

Figure 3.11 Delay from offense to transaction processing.

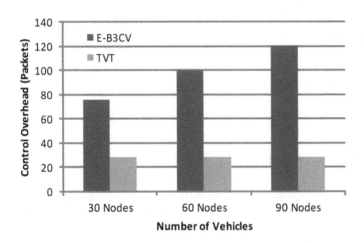

Figure 3.12 Impact of control overhead.

offender. We have designed our approach that vehicles share their records in a cooperative way, which has increased the number of packets in the network.

Figure 3.12 depicts that our proposed work has an increase in the number of packets in the network than TVT as the number of vehicles increase. We have set a percentage of neighbours from the neighbour table who will participate in the process of reporting and validating an offense, which has a direct impact of an increase in packets overhead.

3.5.2.3 Trust in offense processing

Trust in offense processing is defined as how it is ensured to build trust among stakeholders that no query could be raised. In our proposed work, we have involved a set percentage of vehicles for the process of reporting

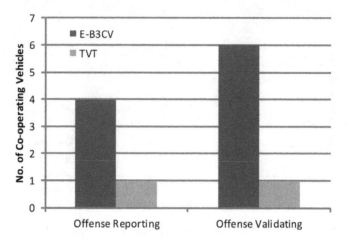

Figure 3.13 Cooperating vehicles for developing trust.

and validating, while in TVT an offender vehicle and CV will inter-communicate to share offenses. Including a number of vehicles ensures that a combined consent for an offense on receiving the offender's packet is taken into account for charging. Figure 3.13 demonstrates a comparison of vehicles involved in offense reporting and validation. The more the number of reporters or validators, the more the trust will be built.

3.5.2.4 Financial gain

Financial gain is defined as the financial impact of resources used in the proposed and competitor technique. In our proposed work there is no extra requirement of hardware or manpower to cover unpatrolled areas because it works with the existing hardware installed, whereas TVT requires a number of CVs patrolling on road with several officers.

Figure 3.14 depicts the calculation of yearly consumption of petrol to cover up a 6 km area of straight road on 26 No., Islamabad, which is shown in Figure 3.15. This is the only calculation of petrol, while the cost of the

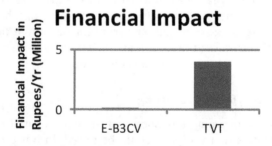

Figure 3.14 Financial impact of resources.

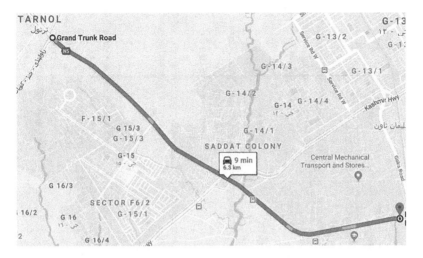

Figure 3.15 Map of 6-km straight road on 26 No., Islamabad.

car and manpower with their expenditures is not included at the moment. By using the existing resources, a minor financial impact with an assumption of maintenance and repair is anticipated.

3.6 CONCLUSION

In this chapter, we have presented an e-challan system for offensive events in the CC-IoV based on Blockchain, which can significantly reduce the mishandling of data resulting from human errors. Also, Blockchain is a well-known technology in dealing with the trust issues among stakeholders, directly or indirectly, involved in the system. Besides developing an immutable and tamper-proof trusted data structure, we have also considered minimising the time taken from offense to the authorities' database. Also, it is necessary to record and issue a ticket for an offense unless it is an offense, and for that we have introduced a validation mechanism which assures that on screening request from the offender, authorities can provide all the sufficient data of an offense. For this validation mechanism, we needed to introduce some extra packets with limited effect over control overhead. We simulated our proposed work in NS-3, based on the provided results performance of our system is better than TVT.

3.7 FUTURE WORK

We have just logged the speeding offense in Blockchain; it can be affected to record all of the auditable activities of road users, without effecting infotainment activities, to minimise human errors, and enhance the trust among stakeholders.

APPENDIX

Table A1.1 Comparison table of literature survey

Techniques	Network	Simulation tool	Strength	Weakness
Blockchain-based public key infrastructure (PKI)	CCN	Not specified	Authentic content distribution	Computational delay
GeoZone	CCVN	ndnSIM	Higher rate of content delivery and lower delay	Static dissemination zone of forwarding interest packet
Density-aware probabilistic interest forwarding	CCVN	ns-2.35	Higher reachability and lower network load	Modified PIT cause processing delay
LolCen	ICVN	Veins (OMNET++, SUMO)	Lower network overhead	Higher delay
Mobility support	IC-IoV based fog computing	ndnSIM	Less consumer delay and drop rate	Lower storage/cache capacity of fog nodes
Smart traffic lights system	VNDN	ndnSIM	Helps in low vision, low power consumption	Limited area coverage by RSUs.
Cognitive route planning	IC-IoV	Practical real-world experimentation	By getting traffic data proposed scheme explore alternative driving routes according to the users requirement based on real time	Lack of user privacy and edge information protection
Data dissemination based on deep learning	CC-IoV	Not specified	Data dissemination rate is high in high traffic density environments along with maximum social score	Data dissemination rate is low in low traffic density environment, in low transmission, rate and in the minimum social score then threshold. High mobility

(Continued)

Table A1.1 (Continued) Comparison table of literature survey

Techniques	Network	Simulation tool	Strength	Weakness
BlockAuth	Blockchain-based ICN	Not specified	Provide more security because of distributed mechanism	Computational overhead increases, required additional storage and number of transmission increases
Decentralised Public Key Infrastructure (DPKI)	Blockchain-based CCN	Not specified	Secure content retrieval and public key tampering is difficult	Computational overhead increases and more storage required
Secure network architecture	Blockchain-based IoVs	Cooja	Provide data integrity, confidentiality	Causes computational and packet overhead
Secure content sharing scheme	Blockchain-based VNDN	Not specified	Secure communication among connected vehicles.	Higher computational overhead
Traffic violation monitoring system	VANETs-based CCN	Not specified	Can control and monitored traffic violations in unpatrolled area	Lack of trust among connected vehicles
Proactive caching	VCCN	ns-3 version 3.2.3	Capable of managing dynamic connection demands that occurs because of mobility	Data store (DS) mobility is missing
Content dissemination	ICN-based VANETs	ns-3	Mitigates broadcast storm and flooding	There is no caching mechanism that lessens the delay and hop count
Caching scheme	CCVN	ONE	Lessens the hops count from content provider along minimum delay	Cache pollution can be done by Sybil attack
Extended architecture	CCN	Omnet++	Minimised the delay by providing data few hops away	Lack of security to avoid evil producers
Mobility supporting	VNDN	Omnet++, SUMO	Provide content fewer hops away and minimises the delay	Lack of trust issues among nodes

(Continued)

Table A1.1 (Continued) Comparison table of literature survey

Techniques	Network	Simulation tool	Strength	Weakness
Semi-automated on-demand insurance	Mobile app installed on customer's vehicle	Not specified	Smart contract-based insurance	Computational overhead increases
Solution to tackle with secure data sharing	Blockchain-based IoVs	Cryptographic library MIRACL	Secure data dissemination among vehicles	Increases network congestion along with delay
Blockchain-based vehicular network architecture	Blockchain-based IoVs	Matlab	Trusted data dissemination among users	Frequently generating blocks occur traffic flooding, lack of validating blocks mechanism
Key management scheme	Blockchain-based NDN	Not specified	Ensured authenticity along with greater trust level on providers content	Time taken authentication, introduces delay
Secure data-sharing mechanism	Blockchain-based CCN	Not specified	Secure data sharing while preserving users privacy	Excessive network usage and delay
Identifier management mechanism	Blockchain-based NDN	Not specified	Secured identifier management	Lack of security mechanism for avoiding malicious nodes
Blockchain-based content delivery	Blockchain-based ICN	Python 3.6	Secure content delivery	Introduces delay in content retrieval
Reputation-based caching	NDN in VANETs	Not specified	Content with low reputation value is not cached	Lack of security mechanism to restrict evil nodes
Integrated Blockchain	VCCN	Not specified	Secured and dynamic connections	Increased computational overhead

Abbreviations: CCN: content-centric networks, CCVN: content-centric vehicular networks, ICVN: information-centric vehicular networks, IC-IoV: information-centric Internet of vehicles, VNDN: vehicular named data networks, VCCN: vehicular content-centric networks, CC-IoV: content-centric Internet of vehicles, ICN: information-centric networks, NDN: named data networks.

Note: Please set as footnote as per the author notes.

Table A1.2 Simulation parameters

Parameter	Value
Operating System	Ubuntu 16.04 LTS
Simulator	Network Simulator-3 (NS3)
Simulation Pattern	SUMO Generated Real-Time Map
Routing Protocol	AODV
MAC/PHY Standard	WIFI_PHY_STANDARD_80211p
MAC Type	Adhoc WiFi MAC
Propagation Loss Model	TwoRayGroundPropagationLossModel
Frequency	5.9 GHz
TxStart Power	20 dBm
TxEnd Power	20 dBm
Packets	VehInfoPkt (8 Bytes), ReportOffPacket (12 Bytes), ValidateOffense (4 Bytes), ReplyValidateOffense (8 Bytes)
No. of Vehicles	30, 60, 90
Simulation Time	300 seconds

REFERENCES

1. H. Khelifi, S. Luo, B. Nour, H. Moungla, Y. Faheem, R. Hussain, et al., 'Named data networking in vehicular ad hoc networks: State-of-the-art and challenges,' *IEEE Communications Surveys & Tutorials*, vol. 22, pp. 320–351, 2019.

2. A. S. T. M. Intl, '*Standard Specification for Telecommunications and Information Exchange Between Roadside and Vehicle Systems-5 GHz Band Dedicated Short Range Communications (DSRC)*,' ASTM E2213-03, West Conshohocken, PA, 2003.

3. D. Jiang and L. Delgrossi, 'IEEE 802.11p: Towards an international standard for wireless access in vehicular environments,' in *VTC Spring 2008-IEEE Vehicular Technology Conference*, pp. 2036–2040.

4. F. Sakiz and S. Sen, 'A survey of attacks and detection mechanisms on intelligent transportation systems: VANETs and IoV,' *Ad Hoc Networks*, vol. 61, pp. 33–50, 2017.

5. R. Gasmi and M. Aliouat, 'Vehicular Ad Hoc NETworks versus Internet of Vehicles—A comparative view,' in *2019 International Conference on Networking and Advanced Systems (ICNAS)*, Annaba, Algeria, pp. 1–6.

6. I. Bhardwaj and S. Khara, 'Research trends in architecture, security, services and applications of Internet of Vehicles (IoV),' in *2018 International Conference on Computing, Power and Communication Technologies (GUCON)*, Greater Noida, India, pp. 91–95.

7. C. V. N. I. Forecast, 'Cisco visual networking index: Forecast and trends, 2017–2022,' *White paper, Cisco Public Information*, 2019.

8. W. Chen, R. K. Guha, T. J. Kwon, J. Lee, and Y.-Y. Hsu, 'A survey and challenges in routing and data dissemination in vehicular ad hoc networks,' *Wireless Communications and Mobile Computing*, vol. 11, pp. 787–795, 2011.

9. M. Amadeo, C. Campolo, and A. Molinaro, 'Enhancing content-centric networking for vehicular environments,' *Computer Networks*, vol. 57, pp. 3222–3234, 2013.

10. C. Bian, T. Zhao, X. Li, and W. Yan, 'Boosting named data networking for data dissemination in urban VANET scenarios,' *Vehicular Communications*, vol. 2, pp. 195–207, 2015.

11. M. Chen, D. O. Mau, Y. Zhang, T. Taleb, and V. C. M. Leung, 'Vendnet: Vehicular named data network,' *Vehicular Communications*, vol. 1, pp. 208–213, 2014.

12. Y. Yu, Y. Li, X. Du, R. Chen, and B. Yang, 'Content protection in named data networking: Challenges and potential solutions,' *IEEE Communications Magazine*, vol. 56, pp. 82–87, 2018.

13. A. Reyna, C. Martín, J. Chen, E. Soler, and M. Díaz, 'On blockchain and its integration with IoT. Challenges and opportunities,' *Future Generation Computer Systems*, vol. 88, pp. 173–190, 2018.

14. N. Kshetri, 'Blockchain's roles in strengthening cybersecurity and protecting privacy,' *Telecommunications Policy*, vol. 41, pp. 1027–1038, 2017.

15. S. H. Ahmed, M. A. Yaqub, S. H. Bouk, and D. Kim, 'Towards content-centric traffic ticketing in VANETs: An application perspective,' in *2015 Seventh International Conference on Ubiquitous and Future Networks*, Sapporo, Japan, pp. 237–239.

16. D. Grewe, M. Wagner, and H. Frey, 'PeRCeIVE: Proactive caching in ICN-based VANETs,' in *2016 IEEE Vehicular Networking Conference (VNC)*, pp. 1–8.

17. Y. Li, X. Su, A. Lindgren, X. Shi, X. Cai, J. Riekki, et al., 'Distance assisted information dissemination with broadcast suppression for ICN-based VANET,' in *International Conference on Internet of Vehicles*, Nadi, Fiji, pp. 179–193.

18. L. Yao, A. Chen, J. Deng, J. Wang, and G. Wu, 'A cooperative caching scheme based on mobility prediction in vehicular content centric networks,' *IEEE Transactions on Vehicular Technology*, vol. 67, pp. 5435–5444, 2017.

19. A. A. Prates, I. V. Bastos, and I. M. Moraes, 'GeoZone: An interest-packet forwarding mechanism based on dissemination zone for content-centric vehicular networks,' *Computers & Electrical Engineering*, vol. 73, pp. 155–166, 2019.

20. A. K. Niari, R. Berangi, and M. Fathy, 'ECCN: An extended CCN architecture to improve data access in vehicular content-centric network,' *The Journal of Supercomputing*, vol. 74, pp. 205–221, 2018.

21. J. M. Duarte, T. Braun, and L. A. Villas, 'MobiVNDN: A distributed framework to support mobility in vehicular named-data networking,' *Ad Hoc Networks*, vol. 82, pp. 77–90, 2019.

22. R. Tizvar and M. Abbaspour, 'A density-aware probabilistic interest forwarding method for content-centric vehicular networks,' *Vehicular Communications*, vol. 23, pp. 100216–100216, 2019.

23. M. Wang, J. Wu, G. Li, J. Li, Q. Li, and S. Wang, 'Toward mobility support for information-centric IoV in smart city using fog computing,' in *2017 IEEE International Conference on Smart Energy Grid Engineering (SEGE)*, pp. 357–361.

24. A. Boukerche and R. W. L. Coutinho, 'LoICen: A novel location-based and information-centric architecture for content distribution in vehicular networks,' *Ad Hoc Networks*, vol. 93, pp. 101899–101899, 2019.

25. M. Al-qutwani and X. Wang, 'Smart traffic lights over vehicular named data networking,' *Information*, vol. 10, pp. 83–83, 2019.

26. C. Zhao, M. Dong, K. Ota, J. Li, and J. Wu, 'Edge-MapReduce-Based intelligent information-centric IoV: Cognitive route planning,' *IEEE Access*, vol. 7, pp. 50549–50560, 2019.

27. A. Gulati, G. S. Aujla, R. Chaudhary, N. Kumar, and M. S. Obaidat, 'Deep learning-based content centric data dissemination scheme for Internet of Vehicles,' in *2018 IEEE International Conference on Communications (ICC)*, pp. 1–6.

28. F. Lamberti, V. Gatteschi, C. Demartini, M. Pelissier, A. Gomez, and V. Santamaria, 'Blockchains can work for car insurance: Using smart contracts and sensors to provide on-demand coverage,' *IEEE Consumer Electronics Magazine*, vol. 7, pp. 72–81, 2018.

29. L. Zhang, M. Luo, J. Li, M. H. Au, K.-K. R. Choo, T. Chen, et al., 'Blockchain based secure data sharing system for Internet of Vehicles: A position paper,' *Vehicular Communications*, vol. 16, pp. 85–93, 2019.

30. T. Jiang, H. Fang, and H. Wang, 'Blockchain-based Internet of Vehicles: Distributed network architecture and performance analysis,' *IEEE Internet of Things Journal*, vol. 6, pp. 4640–4649 2018.

31. S. Sharma, K. K. Ghanshala, and S. Mohan, 'Blockchain-based Internet of Vehicles (IoV): An efficient secure ad hoc vehicular networking architecture,' in *2019 IEEE 2nd 5G World Forum (5GWF)*, pp. 452–457.

32. J. Lou, Q. Zhang, Z. Qi, and K. Lei, 'A blockchain-based key management scheme for named data networking,' in *2018 1st IEEE International Conference on Hot Information-Centric Networking (HotICN)*, pp. 141–146.

33. K. Fan, Y. Ren, Y. Wang, H. Li, and Y. Yang, 'Blockchain-based efficient privacy preserving and data sharing scheme of content-centric network in 5G,' *IET Communications*, vol. 12, pp. 527–532, 2017.

34. H.-K. Yang, H.-J. Cha, and Y.-J. Song, 'Secure identifier management based on Blockchain technology in NDN environment,' *IEEE Access*, vol. 7, pp. 6262–6268, 2018.

35. H. Li, K. Wang, T. Miyazaki, C. Xu, S. Guo, and Y. Sun, 'Trust-enhanced content delivery in blockchain-based information-centric networking,' *IEEE Network*, vol. 33, pp. 183–189 2019.

36. M. Conti, M. Hassan, and C. Lal, 'BlockAuth: BlockChain based distributed producer authentication in ICN,' *Computer Networks*, vol. 164, pp. 106888–106888, 2019.

37. M. Labbi, N. Kannouf, Y. Chahid, M. Benabdellah, and A. Azizi, 'Blockchain-based PKI for content-centric networking,' in *The Proceedings of the Third International Conference on Smart City Applications*, Tetouan, Morocco, pp. 656–667.

38. C. Chen, C. Wang, T. Qiu, N. Lv, and Q. Pei, 'A secure content sharing scheme based on consortium blockchain in vehicular named data networks,' *IEEE Transactions on Industrial Informatics*, vol. 16, pp. 3278–3289 2019.

39. H. Khelifi, S. Luo, B. Nour, H. Moungla, and S. H. Ahmed, 'Reputation-based blockchain for secure NDN caching in vehicular networks,' in *2018 IEEE Conference on Standards for Communications and Networking (CSCN)*, Paris, France, pp. 1–6.
40. V. Ortega, F. Bouchmal, and J. F. Monserrat, 'Trusted 5G vehicular networks: Blockchains and content-centric networking,' *IEEE Vehicular Technology Magazine*, vol. 13, pp. 121–127, 2018.
41. 'Network Simulator—3. Available: www.nsnam.org,' ed.
42. 'Ubuntu 16.04.6 LTS. Available: http://releases.ubuntu.com/16.04,' ed.
43. 'Simulation of Urban Mobility (SUMO). Available: www.dlr.de/ts/en/desktopdefault.aspx/tabid-9883/16931_read-41000,' ed.

Part II

Security and privacy issues in blockchain-enabled fog and edge computing

Chapter 4

Hardware-primitive-based blockchain for IoT in fog and edge computing

Uzair Javaid, Muhammad Naveed Aman, and Biplab Sikdar

CONTENTS

4.1 INTRODUCTION

Security continues to remain one of the key concerns in fog and edge computing. Because edge computing has emerged as an effective offloading mechanism for Internet of Things (IoT) devices, the devices must be made secure to ensure resource availability and computing reliability. With advancements towards urbanisation, smart cities, and a globally connected system enabled by the Internet, cyberattacks and threats are looming large. Because of the increasing complexity of engineering system designs and network architectures, new solutions must be effected.

The need for integrating cloud, fog, and edge infrastructure is consistently being highlighted by the requirement of supporting both latency sensitive and computing intensive IoT applications. For such an integration, it is indispensable that IoT-based environments and their operation be made secure. This chapter primarily focuses on securing IoT environments and the devices associated with them.

4.1.1 Internet of Things

The IoT is one of the most important emerging technologies of the present era aimed at the integration of physical devices in a wide range of applications. Such an integration is made possible with the help of the Internet. The number of IoT devices connected to the Internet in 2016 crossed 6 billion and is expected to reach more than 20 billion by 2020. The devices range from vehicles to bicycles, smart homes to smart cities, closed-circuit television (CCTV) to smart cameras, sensors to radio-frequency identification (RFID) tags, etc. These devices can communicate and share information with each other independent of any human intervention. With the ongoing developments in next-generation system designs, IoT is expected to facilitate and help in resource management, intelligent spaces, smart cities, and industry automation [1,2].

A majority of traditional financial, IoT, and edge computing infrastructures are centralised in nature and rely on third parties for establishing trust in their systems. The third parties may handle accounts and transactions, process and analyse data, and provide security. Such centralised architectures are prone to the following security concerns:

1. *Single point-of-failure:* If a server in a centralised architecture fails, it can potentially disrupt the entire system and render its security services dysfunctional.
2. *Lack of user privacy:* Because centralised systems rely on third parties for their operation, if compromised, they can possibly reveal confidential information or predict data routines. A data routine is a pattern of data usage of a user. Any leaking of such sensitive information may have consequential results [3].
3. *Insecure IoT devices:* An IoT device is an integral part of an edge computing infrastructure. It may be self-sustained and be able to communicate with other devices autonomously while providing constant monitoring and data services. If compromised, it may reveal private and confidential information that may put users at risk.

4. *Limited computational ability:* The IoT devices are small in size. Their processing abilities are resource constrained and that limits them from performing computing intensive tasks.

5. *IP protection:* The IoT devices are user-accessible devices which are vulnerable to intellectual property (IP) theft. Therefore, an IoT device needs to be protected against IP theft, tampering, cloning, and reverse engineering.

6. *Lack of physical protection:* Contemporary security protocols usually make a basic assumption regarding physically protected devices. Although this assumption may be valid for desktops and personal computers (PCs), it is invalid for constrained entities like IoT devices. Therefore, new protocols must be designed to protect IoT devices from physical attacks.

Figure 4.1 presents an overview of a traditional IoT network where different kinds of sensors and devices interact with a central server through a communication channel. It is worth noting here that cyber threats exist from device (physical) level to communication and network levels. Moreover, the resource-constrained nature of IoT devices exacerbates the security challenges [1,5,6].

Figure 4.1 A conventional Internet of Things (IoT) architecture. Proposed by [4] (From Javaid, U. et al., BlockPro: Blockchain Based Data Provenance and Integrity for Secure IoT Environments, in *Proceedings of the 1st Workshop on Blockchain-enabled Networked Sensor Systems*, BlockSys'18. Shenzhen, ACM, China, 13–18, 2018.)

This makes it impractical to apply classical security techniques to constrained IoT devices. Therefore, new protocols and frameworks are needed.

This chapter proposes the use of physical unclonable functions (PUFs) with blockchain and smart contracts for establishing digital provenance and preserving data integrity. A PUF can be formally described as a system that maps a set of challenges to a set of responses based on the physical microstructure of a device. This way, it makes it nearly impossible to modify, clone, or tamper with a PUF [7], thereby providing a unique hardware fingerprint for each IoT device (see Section 1.3).

4.1.2 Blockchain

Blockchain is an online distributed ledger of cryptographically secured blocks [8]. A block is an instance that collates tuples of transactions together with respect to a specific time interval. The first block in a blockchain is called the 'genesis' block. All the succeeding blocks are added to it in a chronological order. The genesis block B_i is hashed and stored in the second block B_{i+1}. The hash of B_{i+1} is stored into B_{i+2}, and so on. This way, each block B_j has a hash of the previous block B_{j-1}. This forms the basis of the blockchain and can simply be formulated as:

$$B = \left(B_0, B_1, B_2, \ldots\right) \tag{4.1}$$

$$B_i = \sum_{j=0}^{m} \left(tx, \; payload\right)_{ij} \tag{4.2}$$

where B represents a blockchain and B the chronologically appended blocks in it. Payload here represents data of concern with application specific information. For reference, the index 0 is taken as the genesis and first block. From Equation 4.2, a block can contain m instances of transactions, data, and other information depending on the nature of the application it is designed for.

Blockchain operates in a peer-to-peer (P2P) fashion (i.e., it depends on its network constituents for its resources and computation). There is no central authority here, but a collective mechanism is employed over its constituents: The nodes. This is so that system nodes can work together and secure the blockchain from adversaries. Although double spending has been a problem for a long time in both centralised and decentralised payment protocols, the blockchain makes it infeasible unless and until an adversary or a group of adversaries gain control of 51% computational power of the network [9].

After the genesis block, the succeeding blocks have to be mined by miners. Mining is the process of adding new blocks to the blockchain, thereby increasing its size. Miners are volunteers who provide their resources for the blockchain to operate. For each block to be mined, the miners have to solve a cryptographical puzzle and find a desired target. They do so by following a distributed consensus algorithm, proof-of-work.

The most widespread use of blockchains is in cryptocurrencies such as Bitcoin [10]. However, many applications have adopted blockchains to provide decentralised and trust-free solutions [11]. Another use of blockchains is with computer programs (*smart contracts*) such as Ethereum [12]. One of the advantages of using blockchains with IoT environments is that it allows IoT devices to transact freely without relying on third parties. However, scalability along with standardisation in IoT and blockchains remain key developmental concerns [13].

4.1.2.1 Transaction

A transaction, *tx*, is a set of instructions which changes the ownership of tokens or digital assets from one user to another. A token is a virtually valued currency. The ownership of these tokens or assets is changed by digitally signing a transaction which contains these tokens or assets with a unique private key and then broadcasting it in the network. The network can then verify this transaction signature using the corresponding public key. Mathematically, a transaction can be formulated as:

$$tx = t_{in}Num\|t_{in}\|t_{out}Num\|t_{out}\|nonce\|data\|t \qquad (4.3)$$

where t_{in} and t_{out} are the input and output vectors of a transaction that represent both the sender and recipient. These consist of tuples of elements and are used when the ownership of tokens is digitally transferred from one user to another one. Here, $t_{in}Num$ and $t_{out}Num$ are the total number of incoming and outgoing transactions relative to a timestamp t, respectively. The *nonce* field contains the challenge for miners to mine the transaction and validate it, whereas the *data* field contains information.

All transactions relative to a timestamp are collated together which form tuples of transactions to be mined. When a transaction occurs and is broadcast in the network successfully, the ownership of tokens or assets changes and the new owner is announced by the sender. The network participants trace the transaction history and verify its signature before mining.

4.1.2.2 Signing a transaction

Before a transaction, *tx*, can be stored, it has to be validated to prove its authenticity. Authenticity in a *tx* is achieved by digitally signing it with a private key for encrypting and then validating it by decrypting with the corresponding public key. For a user, *i*, it can mathematically be formulated as:

$$sign_i = f(private_{key_i}, payload) \tag{4.4}$$

$$verify_i = f(public_{key_i}, payload) \tag{4.5}$$

where *sign*$_i$ represents the signature of user *i* and is a function of *private*$_{key_i}$ and *payload*, which are the private key of and data to be signed by user *i*, respectively. Similarly, *verify* is a function of *public*$_{key_i}$ and the same payload. The only way a user can verify the signature *sign*$_i$ is with the help of the corresponding public key of user *i*.

4.1.2.3 Reaching consensus: Proof-of-Work

The proof-of-work (PoW) is a distributed consensus mechanism whose main function is to present a puzzle (an exceptionally difficult mathematical problem) to miners each time a block is verified. The puzzle has a target value, *v*, which is a 256-bit number with specific *n* number of zeros as significant digits. The choice of *n* defines the difficulty of the puzzle. The PoW is an exponential function, which means a lower target value will require more attempts of hashing. The solution finding process is a probabilistic one because to find the target value, a miner has to change its hashing iteratively. This is because each input produces a unique hash and different hashing attempts are required to solve the challenge. Blockchain makes the double-spending attacks infeasible because of this attribute.

For example, the PoW function of Bitcoin can be formulated as:

$$pr(h \leq \upsilon) = \upsilon / 2^{256} \tag{4.6}$$

where the probability of finding a proof (mining) *h* for a target υ is $pr(h \leq \upsilon)$. When a miner finds the target value, it broadcasts its proof in the network. The other miners then rerun this proof and validate the block. Once the block is validated and accepted by the majority of the network, it is then finally added to the blockchain.

4.1.3 Physical unclonable functions

A physical unclonable function (PUF) is a 'hardware fingerprint' that can provide semiconductor devices (such as microprocessors, integrated circuits [ICs], etc.) with unique identities. PUFs are used in place of secret keys or passwords that can be used to assign a hardware fingerprint to

IoT devices [14]. They are characterised by and are based on the variations that naturally occur during the manufacturing process of semiconductor devices. This provides a way for differentiating between otherwise identical semiconductor devices. PUFs are generally used in cryptography and for applications with high-security requirements.

The PUFs provide a device with distinctive and unique hardware fingerprints determined by their responses to challenges. For example, if a challenge is input to a PUF many times, it will always produce the same output (response) with a high probability. Similarly, in a different PUF, if the same challenge is given as an input, it will always produce a different output with high probability [15]. This way, PUFs provide physical security to devices with unique fingerprints that cannot be altered or tampered with. A common example of a PUF is the subscriber identification module (SIM) card used in cellular networks.

The establishment of data provenance in this chapter is based on PUFs. A PUF is characterised by a challenge-response pair (CRP) which can be represented as:

$$R^i = pr(C^i). \tag{4.7}$$

where R^i is the response generated by a challenge C^i.

4.2 PUF-BASED BLOCKCHAIN PROTOCOLS FOR IOT IN FOG AND EDGE COMPUTING

Traditional IoT environments involve varying security features and rely on the computational capabilities of devices, servers, and other components associated with the network. The IoT devices are prone to physical attacks and can be hacked easily because their low-level architecture. Moreover, preserving data integrity and securely transmitting data in IoT environments has also proven to be vulnerable to attacks because they can be easily hacked and tampered with once an adversary gains access. This chapter integrates PUFs with blockchain for a safe and secure IoT environment that not only ensures data provenance but also enforces data integrity by providing an immutable storage platform. This framework provides the following security features:

1. A hardware primitive-based blockchain for blocking unregistered devices, establishing data provenance and preserving data integrity.
2. A decentralised network for control and trust-free operation of IoT environments.
3. Unique hardware fingerprints for IoT devices.

4.2.1 Data provenance protocols

The existing literature on establishing data provenance using hardware primitives in IoT systems is limited, especially when it comes to protocols that employ blockchain and PUF. A blockchain-based protocol, BlockCloud, for data provenance in cloud computing was proposed by [16]. Their system design uses blockchain with proof-of-stake (PoS) consensus mechanism and has two verification processes to establish data provenance: the transaction verification prior to insertion in the block and the verification of a block to evaluate for data tampering before updating the blockchain with the consensus winners block. Although they establish robust defence mechanisms against a number of attacks while preserving data integrity, they fail to defend against physical attacks on constituents of a cloud computing system.

Furthermore, sensor PUF was proposed by [17]. Unlike normal PUFs, the response of sensor PUFs is based on not only challenge but also a sensed quantity. Thus, a sensor PUF is characterised by a challenge-quantity-response instead of a CRP, thereby providing authentication, unclonability, and verification of a sensed value. The sensor PUF was proposed to solve the problem of spoofed measurements in which an attacker tampers with the analogue signals that go from the sensor entity to the embedded microcontroller. Therefore, sensor PUFs can be used to verify the integrity and establish data provenance for a specific sensor.

This chapter discusses a protocol proposed by [4], where a PUF-based blockchain protocol is used to establish data provenance and preserve data integrity in IoT-based environments. The proposed protocol consists of two phases: A setup phase for IoT device registration and a data transfer phase for communication between the devices.

4.2.1.1 The setup phase

The setup phase is enabled by smart contracts in the blockchain network. The contract provides secure communication among IoT devices and the blockchain network users. Figure 4.2 holistically illustrates the information flow layout of IoT devices and the blockchain network. The devices interact with the contract, which in turn interacts with the desired user or function of the network. Before data can be transferred by a device, it has to be registered first. The functions *reg.iot device(addr)* and *del.iot device(addr)* are responsible for registering and deleting devices with respect to their IP addresses, respectively. Moreover, *list.reg iot devices* maintains a list of all trusted devices registered in the network. The *CR.iot dev PUF* maintains a list of PUF CRPs for the registered devices.

The contract can be coded using Solidity; an illustration of its interface can be seen in Figure 4.3. Moreover, the operation of the contract consists of two phases:

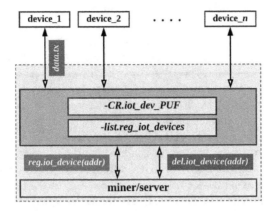

Figure 4.2 The information flow layout of setup phase. (From Javaid, U. et al., BlockPro: Blockchain Based Data Provenance and Integrity for Secure IoT Environments, in *Proceedings of the 1st Workshop on Blockchain-enabled Networked Sensor Systems*, BlockSys'18. Shenzhen, ACM, China, 13–18, 2018.)

Compile   Run   Settings   Analysis   Debugger   Support

Environment	Injected Web3	Rinkeby (4) ▼ i
Account	0xa0f...97361 (2.979562348 ether)	▼
Gas limit	3000000	
Value	0	wei ▼

Register ▼

Deploy

Load contract from Address   At Address

Transactions recorded: ①   ⌄

Deployed Contracts   🗑

▸   Register at 0xd42...ac7ba (blockchain)   ✕

Figure 4.3 User interface of Solidity IDE.

1. *Initialise*: For initialisation of the contract, a *server* node (e.g., operated by the owner of an IoT device or a cloud-based service provider) deploys the smart contract. This will allow the device/provider to be recognised as the *server* variable by the contract. After the initialisation of the contract, the IP address of the contract will be broadcast in the network so that devices and users can interact with it.

2. *Interact*: In this phase, IoT devices interact with the contract to get registered as the contract provides device registration. For registration, a device first registers its PUF-CRP in the network. This CRP is stored by the contract and is used for establishing data provenance. Moreover, along with their PUF-CRP, the IP address of devices is also stored by the contract. When a device transmits some data, the contract checks if it is in the registered list of devices. If it is not in the registered list, the request is terminated. Otherwise, a PUF challenge is sent to the device; if it generates a positive response to this challenge, the request is entertained.

4.2.1.2 The data transfer phase

Once the devices are registered, they need to interact with other devices or users in the network. This requires them to transfer data and requests. For doing so, they must pass their PUF-CRP test. This chapter assumes that each IoT device comes equipped with a PUF and the response to a specific challenge can be obtained using only two ways (i.e., either by the device using its PUF or by the operator from a saved copy in its memory). When a certain device, d_i, is to be registered, a CRP for its PUF is already recorded by the operator in the network using the smart contract. This way, each device has its own unique ID along with a unique CRP stored in the contract.

Data provenance is achieved using PUFs. After a request is generated by a device, d_i, the contract checks its validation using the algorithm detailed in Algorithm 4.1. In this algorithm, function $\mathrm{PUF}_{CR}(d_i)$ is used to check provenance of data from device d_i. When d_i transmits data, the algorithm first checks if the data is coming from a trusted device. If it is registered, then the algorithm checks whether its PUF challenge-response is valid or not. It does so by invoking the PUF challenge-response protocol shown in Figure 4.4. The steps for this protocol are as follows:

Algorithm 4.1:   PUF challenge-response validation

---

1   function: $\mathrm{PUF}_{CR}(d_i)$
    Input : $\mathrm{tx}(d_i)$
    Output: *pass*, *fail*
2   if *(tx(d_i) is uploaded* **and** *tx(d_i) is valid)* then
        // Check d_i is registered/not

```
3    if (d_i is in the trusted list) then
        // Check d_i has positive PUFd_i
        // invoke PUF challenge-response protocol
4        if (PUFd_i response = true) then
5            return pass
6        else
7            return fail
8        end
9    else
10        return fail
11    end
12    else
13    return fail
14    end
15    end function
```

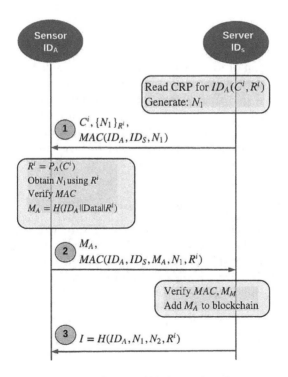

Figure 4.4 The physical unclonable function (PUF)-based challenge-response protocol.

1. The server in the network with identity ID_S reads the CRP (C^i, R^i) for device ID_A and generates a nonce N_1 for it.
2. The server ID_S then sends the nonce N_1 which is encrypted using R (i.e., $\{N_1\}$ R^i and the challenge C^i to the IoT device ID_A in message 1).
3. Upon obtaining the nonce from the server ID_S, IoT device ID_A then obtains the corresponding response R^i for the challenge C^i with the help of its PUF.
4. After obtaining the response R^i, ID_A performs the following steps:
   a. Obtain N_1 using R^i as the secret key. Generate a random nonce N_2.
   b. Verify and validate the message authentication code (MAC) using the parameters in its memory to preserve data integrity.
   c. Once it verifies the MAC, its produces a hash: $H(ID_A, data, N_1, N_2, R^i)$ and sends it to the server in message 2.
5. Once the server ID_S receives message 2 from IoT device ID_A, it checks and verifies the MAC and hash. If both are valid, the request from the device is entertained. Otherwise, the request is dropped. The server then sends an acknowledgement to the IoT device ID_A in the form of authentication parameter $I = H(ID_A, N_1, N_2, R^i)$.

4.2.2 Performance analysis

By using a hardware-primitive-based blockchain, a decentralised and trust-free operation of IoT devices is obtained. It is able to provide defence against impersonation and data-tampering attacks. Instead of the conventional centralised IoT architecture having a single server as illustrated in Figure 4.1, blockchain has decentralised architecture. Moreover, the ability to deploy smart contracts ensures that the devices operating are trusted ones. The advantages of using such an architecture can be reflected in the ways described in the following sections.

4.2.2.1 Centralised versus decentralised architecture

Figure 4.1 shows a traditional, centralised IoT architecture with its security shortcomings. Let us consider n IoT devices, each having a normalised resource requirement of r_i. If the maximum number of available servers is S with available resources r, then we can mathematically formulate a conventional IoT architecture as:

$$\sum_{i=1}^{n} r_i \leq S_r \qquad (4.8)$$

A centralised architecture design is prone to a single point-of-failure that can possibly bring down the whole system. Moreover, computations are not distributed but mainly concentrated in a centralised fashion in the network. This chapter addresses these issues by providing a decentralised platform. By using blockchain as a platform, the computation is distributed among the constituents of the network. This eliminates single-point-of-failure, it empowers the system to withstand them, and to continue operating even if a number of constituents cease to function.

4.2.2.2 Data provenance and data integrity

Hardware-primitive-based blockchain uses PUFs to validate the source of the data. As each PUF produces a unique response, therefore, data provenance is established through the use of PUFs with each IoT device. In addition, the use of the blockchain platform enforces data integrity. Blockchain provides an immutable chain of records (i.e., starting from the first block, the subsequent ones are added in a chronological order). To change one block, one must trace it back to the first one which is practically infeasible. By using a hardware-primitive-based blockchain, the following salient features can be derived:

1. Each IoT device has a unique ID relative to its PUF. This provides immunity from impersonation attacks, thereby providing a way for establishing data provenance.
2. The smart contracts provide a safe and secure mechanism for the transmission, authentication and storage of requests and data, respectively.
3. With an ever-growing chain of records, all the data is validated first, and then collated together to form a block relative to a specific timestamp. Then the block is verified by the miners and stored on the blockchain permanently, which cannot be tampered with afterward, thereby providing and preserving data integrity.

4.2.2.3 Distributed consensus

Traditional IoT system design relies profusely on trust because it is one of the enabling factors of a system's operation. Typically, because of centralised structures, there is a third party involved between an IoT device and a server. This third party may be a storage solution, an entity providing computational power or other forms of service. Although the notion of service is lucrative, it does not come for free. The inclusion of third parties involves extra time and labour along with an associated monetary cost.

Hardware-primitive-based blockchain eliminates the need for third parties by distributing computation and consensus among the participants of the network. Not only are they responsible for providing the necessary computational power for the network to operate, but they also provide a trust-free environment using distributed consensus protocols. The distributed consensus protocol used in this chapter is PoW. This cuts down the extra cost, time, and labour associated with third parties and puts the control back into the hands of the network constituents.

4.2.3 Security analysis

This section presents a formal security analysis of this chapter. The following set of assumptions is made and a threat model described.

4.2.3.1 Assumptions

1. Every IoT device is equipped with a PUF.
2. The PUF and IoT device form a system on chip (SoC) and any kind of tampering will render the PUF useless [18,19].
3. The PUF and microcontroller communicate over a secure channel given the SoC assumption [18,19].
4. The standard assumption regarding a PUF: Every PUF is unique and unclonable (i.e., an adversary cannot predict its behaviour [20]). A PUF can be modelled as PUF:$\{0,1\}^{l_1} \rightarrow \{0,1\}^{l_2}$, that is, a PUF will produce an output of length l_2 when excited with an input of length l_1. This chapter models PUF security using a security game, $\text{Exp}_{PUF,A}^{Sec}$, between a challenger, C, and adversary, A:
   a. A randomly chooses a challenge C^i and sends it to C.
   b. C uses the PUF to obtain the response R^i and reveals R^i to A.
   c. C selects a random challenge C^x, which has not been used before and obtains the response R^x using the PUF, that is, $R^x = PUF(C^x)$.
   d. A can query the PUF a polynomial number of times for challenges other than C^x.
   e. A outputs its guess $R^{x'}$ for the challenge C^x.
   f. A wins the game if $R^{x'} = R^x$

The advantage of the adversary A in this game can be modelled by $\text{Adv}_A^{PUF} = \Pr\left[R^{x'} = R^x\right]$.

4.2.3.2 Threat model

A set of IoT devices $M = M_1, M_2, \dots , M_n$ interact with the secure blockchain S. The devices communicate with S over an unsecured network. At the conclusion of the authentication phase, the entities are either registered in the network or rejected. If authentication is successful, a device can start to transmit data by interacting with the smart contract. The adversary A is assumed to have full control over the communication channel between devices and the miners in the blockchain. This may include attacks like eavesdropping, tampering, replaying, and injecting packets in the network. The following set of queries is used to model these attacks:

- SendS(S, m0,r0,m1) is used to model the query where *A* acts like a legitimate device and sends a message m0 to *S* and receives r0. The device then replies to *S* with m1.
- SendM (ID, m0,r0) is used to model the query where *A* acts like a blockchain and sends a message m0 to a device and receives r0.
- Monitor(M, S) models *A*'s ability to continuously eavesdrop on the radio channel between device *M* and *S*.
- Drop(A) models *A*'s ability to drop packets between *M* and *S*.

The adversary *A* can call SendS, SendM, Monitor, and Drop any polynomial number of times.

4.2.3.3 Security proofs

Lemma 1. *It is not possible for an adversary to tamper with the data in a blockchain.*

Proof. A blockchain is composed of chronological blocks hashed together starting from the genesis block up until the latest block. Therefore, to tamper with the data in a single block, an adversary needs to successfully redo the PoW for that block and all the preceding blocks in the blockchain as well. However, to achieve this the adversary needs to have at least 51% of the total computational power of the blockchain network [10]. Given a decent-sized blockchain network, such attacks are extremely difficult or even impossible.

Lemma 2. *It is not possible to reveal the secret response of an IoT device.*

Proof. Every IoT device has its own PUF. During the authentication phase, a miner sends a challenge C^i to the device, and the device uses this challenge to generate the secret response, R^i. Thus, the device does not store the secret response R^i in its memory and only generates it when needed. Therefore, even if an adversary launches a physical attack on a IoT device, he or she cannot obtain the secret response R^i. This shows that an adversary has no possible way of extracting or revealing the secret response for a smart meter.

Theorem 3. *PUFs achieves mutual authentication of an IoT device and blockchain.*

Proof. An adversary *A* may try to authenticate itself as a legitimate IoT device. We can model this by the following game between a challenger *C* and adversary *A*.

1. *C* selects a legitimate device, M_1, and registers it with a miner in the blockchain.
2. *A* calls SendS, SendM, Monitor, and Drop a polynomial number of times on the miner and the device M_1.

3. A invokes the SendS oracle to authenticate itself as a legitimate device to the miner.
4. A wins the game if he or she can successfully complete the authentication phase.

In the authentication phase, the IoT device needs to generate the secret response R^i to successfully authenticate itself with the miner, that is, to pass the verification process, the device needs to successfully create the authentication parameter $MAC(ID_A, ID_S, M_A, N_1, R^i)$. Therefore, when A attempts to authenticate itself with the miner, he or she also needs to produce a valid authentication parameter. However, to construct a valid authentication parameter, the adversary needs the secret response, R^i. By lemma 2, this is not possible. Thus, an adversary cannot successfully authenticate itself as a legitimate IoT device.

The second part of the proof considers the case when an adversary A attempts at acting as a miner and fooling a device into authenticating it as the miner. This can be modelled by a security game similar to the previous one except that in step 3, instead of calling the SendS oracle, A calls SendM to impersonate the legitimate miner. To successfully impersonate the miner, the adversary needs to send a valid authentication parameter $MAC(ID_A, ID_S, N_1, R^i)$ to the device. However, by lemma 2, A cannot obtain R^i. Thus, we can conclude that successful authentication of the miner and the device is achieved.

Theorem 4. *Data provenance: The proposed protocol successfully establishes the authenticity of the origin of data.*

Proof. An adversary may try to impersonate a legitimate IoT device and send tampered data to a miner. We can model this by the following security game:

1. C selects a device M_1 and uses it to perform a transaction.
2. A calls SendS, SendM, Monitor, and Drop a polynomial number of times on the miner and the device.
3. A calls the SendS oracle to impersonate a smart meter.
4. If A can successfully authenticate the tampered data sent by it to the miner then A wins the game.

To prove his or her legitimacy and successfully tamper with the data of IoT devices, A has two options: first, tamper with the blockchain and second, tamper with the hashed data parameter in message 2 during the authentication phase (i.e., $M_m = H(ID_A \| Data \| R^i)$). However, by lemma 1, tampering with the blockchain is not possible. Moreover, by lemma 2, the adversary cannot obtain R^i and thus, cannot tamper with the hashed data parameter. This shows that data provenance is achieved and data tampering attacks are avoided.

Lemma 5. *PUFs are safe against physical and cloning attacks.*

Proof. Physical attacks can be used by an adversary to extract secret keys from the memory of a IoT device. However, as shown in lemma 2, IoT devices do not store the secret response R^i (used to establish the various security properties) in their memory. Moreover, because of the SoC assumption, the PUF cannot be separated from a device and neither can an adversary eavesdrop on the communication between the PUF and the IoT device. This provides a defence mechanism against physical attacks.

4.2.3.4 Formal proofs

This section presents the formal security proofs of the proposed protocol using the Mao and Boyd logic [21]. Such formal poofs are important to establish the security properties of a protocol such as secrecy, and authentication among others. The BAN logic is commonly used for the security analysis of security protocols [22]. However, several weaknesses in the BAN logic were identified by [21]. Therefore, this section uses the Mao and Boyd logic for the security proofs, which is an improved extension of the BAN logic. For ease of notation, we represent IoT device ID_A with A, the server with S, and data sent by the IoT device with Δ.

The first step in analysing a protocol using the Mao and Boyd logic is to idealise the protocol messages. The details on protocol message idealisation are given in the Appendix at the end of this chapter. The idealised version of the messages exchanged during the PUF challenge-response protocol during the data transfer phase is as follows:

1. $S \rightarrow A : A, \{N_1\}_{R^i}$.
2. $A \rightarrow S : \{A, \Delta RN_1 RN_2\}_{R^i}$.
3. $S \rightarrow A : \{A, N_1 RN_2\}_{R^i}$.

A prerequisite of the Mao and Boyd logic is an effective way for data integrity and origin verification. The proposed protocol uses MACs and hash operations to establish data integrity and origin verification. The Mao and Boyd logic depends on a set of inference rules to prove security properties. The set of inference rules used in this analysis can be found in the Appendix. The set of initial beliefs for the proposed protocol are as follows:

1. $A \models A \overset{R^i}{\leftrightarrow} S$ and $S \models A \overset{R^i}{\leftrightarrow} S$: S saves a CRP for each IoT device in its memory, while A can generate R^i by using the respective challenge.
2. $A \models S^c \triangleleft \|N_2$ and $S \models A \models \{S\}^c \triangleleft \|N_2$: N_2 is generated by A.
3. $A \overset{R^i}{\sim} N_2$: Message 2 in the idealised protocol.
4. $A \models \#(N_2)$: A generates a new N_2 each time.

5. $S \models \#(N_1)$: S generates a new N_1 each time.

6. $A \models \sup(S)$: S is the super principal with respect to N_1.

7. $S \models \sup(A)$: A is the super principal with respect to N_2.

8. $A \overset{R^i}{\triangleleft} N_1$ R N_2: Message 3 in the idealised protocol.

9. $S \overset{R^i}{\triangleleft} N_1$ R N_2: Message 2 in the idealised protocol.

10. $A \models S \models \{A\}^c \triangleleft \|N_1$ and $S \models A^c \triangleleft \|N_1$: S generates a new N_1 each time.

11. $S \overset{R^i}{\mid\sim} N_1$: Message 1 in the idealised protocol.

12. $A \overset{R^i}{\mid\sim} N_2$: Message 2 in the idealised protocol.

13. $A \models \#(\Delta)$: A generates the data.

14. $A \overset{R^i}{\mid\sim} \Delta$: Message 2 in the idealised protocol.

15. $A \models S^c \triangleleft \|\Delta$: A generates the data.

16. $S \overset{R^i}{\triangleleft} N_1$ R Δ: Message 2 in the idealised protocol.

The tableau of Figure 4.5, proves the fact that A believes N_2 is a good shared secret between A and S (i.e., $A \models A \overset{N_A}{\leftrightarrow} S$). The good-key rule from (5) can now be applied. This rule states that N_2 is a good secret between A and S, if A is certain that no one else except A and S has seen $N_2 (A \models \{A,S\}^c \triangleleft \|N_2)$ and that N_2 is a fresh nonce $(A \models \#(N_2))$. In turn, the confidentiality rule from (2) can be used to prove $A \models \{A,S\}^c \triangleleft \|N_A$, for which it needs to be shown that A and S share a good secret $R^i (A \models A \overset{R^i}{\leftrightarrow} S)$, and A sent N_2 to S encrypting it with R^i $(A \overset{R^i}{\mid\sim} N_2)$. We can find these statements as well as $A \models \#(N_2)$ in the set of initial beliefs. Thus, the initial statement (i.e., $A \models A \overset{N_A}{\leftrightarrow} S$) is proved. Similarly, Figure 4.6 establishes the fact that N_1 is a good shared secret between A and S. The tableaux of Figures 4.7 and 4.8 shown the secrecy proofs N_2 and N_1 on the site of principal S, respectively. These tableaux also prove the authentication properties [20]. To prove provenance of the data, a similar analysis is done for the data Δ as given in the tableaux of Figures 4.9 and 4.10, respectively.

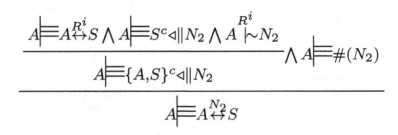

Figure 4.5 Proof of 'A believes N_2 is a good shared key of A and S'.

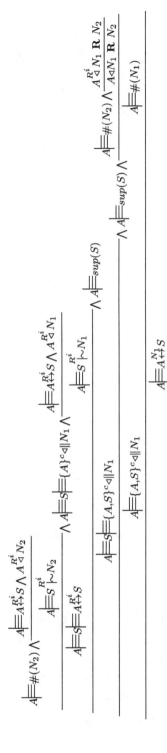

Figure 4.6 Proof of 'A believes N_1 is a good shared key of A and S'.

$$\frac{S\models A\overset{R^i}{\leftrightarrow}S \wedge S\models A^c\triangleleft\|N_1 \wedge S\overset{R^i}{\mid\!\sim}N_1}{S\models\{A,S\}^c\triangleleft\|N_1} \wedge S\models\#(N_1)$$

$$\frac{}{S\models A\overset{N_1}{\leftrightarrow}S}$$

Figure 4.7 Proof of '**S** believes N_1 is a good shared key of **A** and **S**'.

$$S\models\#(N_1)\wedge\frac{\dfrac{S\models A\overset{R^i}{\leftrightarrow}S\wedge S\overset{R^i}{\triangleleft}N_1}{S\models A\overset{R^i}{\mid\!\sim}N_1}}{S\models A\models A\overset{R^i}{\leftrightarrow}S}\wedge S\models A\models\{S\}^c\triangleleft\|N_2\wedge\frac{S\models A\overset{R^i}{\leftrightarrow}S\wedge S\overset{R^i}{\triangleleft}N_2}{S\models A\overset{R^i}{\mid\!\sim}N_2}\wedge S\models_{sup}(A)\qquad S\models\#(N_1)\wedge\frac{S\overset{R^i}{\triangleleft}N_1\text{ R }N_2}{S\triangleleft N_1\text{ R }N_2}$$

$$\frac{S\models A\models\{A,S\}^c\triangleleft\|N_2}{S\models\{A,S\}^c\triangleleft\|N_2}\wedge\qquad S\models\#(N_2)$$

$$\frac{}{S\models A\overset{N_2}{\leftrightarrow}S}$$

Figure 4.8 Proof of '**S** believes N_2 is a good shared key of **A** and **S**'.

$$S\models\#(N_1)\wedge\frac{\dfrac{S\models A\overset{R^i}{\leftrightarrow}S\wedge S\overset{R^i}{\triangleleft}N_1}{S\models A\overset{R^i}{\mid\!\sim}N_1}}{S\models A\models A\overset{R^i}{\leftrightarrow}S}\wedge S\models A\models\{S\}^c\triangleleft\|\Delta\wedge\frac{S\models A\overset{R^i}{\leftrightarrow}S\wedge S\overset{R^i}{\triangleleft}\Delta}{S\models A\overset{R^i}{\mid\!\sim}\Delta}\wedge S\models_{sup}(A)\qquad S\models\#(N_1)\wedge\frac{S\overset{R^i}{\triangleleft}N_1\text{ R }\Delta}{S\triangleleft N_1\text{ R }\Delta}$$

$$\frac{S\models A\models\{A,S\}^c\triangleleft\|\Delta}{S\models\{A,S\}^c\triangleleft\|\Delta}\wedge\qquad S\models\#(\Delta)$$

$$\frac{}{S\models A\overset{\Delta}{\leftrightarrow}S}$$

Figure 4.9 Proof of '**S** believes Δ is a good secret of **A** and **S**'.

$$\frac{A\models A\overset{R^i}{\leftrightarrow}S \wedge A\models S^c\triangleleft\|\Delta \wedge A\overset{R^i}{\mid\!\sim}\Delta}{A\models\{A,S\}^c\triangleleft\|\Delta} \wedge A\models\#(\Delta)$$

$$\frac{}{A\models A\overset{\Delta}{\leftrightarrow}S}$$

Figure 4.10 Proof of '**A** believes Δ is a good secret of **A** and **S**'.

4.3 CONCLUSION

The IoT will undoubtedly be globally pervasive in the near future and will play a key role in the advancement of fog and edge computing. Although IoT demonstrates potential benefits for fog, edge, and cloud infrastructure integration, one of the key concerns is its security (i.e., how to secure an IoT environment and its operation). Traditional security techniques are inapplicable to a wide range of IoT devices because of their resource constraints and hardware limitations (i.e., these protocols focus on network-level threats and ignore hardware-level ones). Therefore, new protocols must be proposed to secure IoT devices. This chapter presented a hardware-primitive-based blockchain for establishing data provenance and preserving data integrity in an IoT environment. The hardware primitive used is PUF. The use of PUFs gives each IoT device a unique hardware fingerprint that is exploited for establishing provenance. Moreover, the decentralised paradigm for data storage and retrieval using blockchain forms the basis for preserving integrity. By using PUF-based blockchain with smart contracts, a defence mechanism against impersonation and tampering attacks can be realised. The security analysis demonstrated that the hardware-primitive-based blockchain security protocols are not only secure against network-level threats but are also resilient to physical attacks. However, a concern in blockchain technology lies in its ability to scale. The size of the chain in blockchain is bound to grow over time; how to effectively as well as efficiently scale it is an open research question. One way of addressing this issue is using directed acyclic graph (DAG), such as IOTA.[1] IOTA uses a DAG-based blockchain in which new incoming transactions can verify other transactions at any time. It uses *Tangle* [22] and its operation is relative to the arrival rate of incoming transactions, (i.e., higher number of new transactions will create lower verification times and higher throughput). The idea looks promising that can be exploited to develop a scalable blockchain solution for IoT and help integrate fog, edge, and cloud computing infrastructures together.

APPENDIX

A brief introduction to the formal analysis of the proposed security protocols is presented in this section using the Mao and Boyd logic [20]. To understand their technique, the protocol messages need to be idealised first. Three types of information are used to construct logical formulae: M: messages, P: principals, and F: formulae. The notation of capital letters A, B,

---

[1] https://www.iota.org

P, Q, ... is used to represent principals, letters K, M, N, ... to denote messages, and X, Y, Z, ... for formulae. For analysis, the following predicate constructs are used:

- $P \models X$: P believes X is true and may act accordingly.
- $P \overset{K}{\sim} X$: P encrypted X using key K.
- $P \overset{K}{\triangleleft} X$: P sees X using decipherment key K. In the absence of encryption we use $P \triangleleft X$.
- $P \overset{K}{\leftrightarrow} Q$: K is a good shared key for P and Q.
- $\#(M)$: M is fresh (not used before).
- $\sup(S)$: Principal S is the trusted party.
- $P \triangleleft \| M$: Message M is not available to principal P.

Furthermore, the following definitions are presented to better understand the rules for protocol messages idealisation.

- *Atomic Message*: A piece of data in a message constructed without using any of the symbols ',', 'I', 'R', or '{}' is called an atomic message. The symbol ',' is used to separate fields in a message and '{}' for encryption. The purpose of the symbols 'I' and 'R' is defined below.
- *Challenge*: An atomic message sent and received in separate lines by the same principal (the originator). A timestamp is not considered an atomic message.
- *Replied Challenge*: A challenge that appears in a message sent to the originator.
- *Response*: An atomic message (except timestamps) and a replied challenge sent together by the sender of the response.
- *Nonsense*: An atomic message is a nonsense if it is not a challenge, response, or a timestamp.

The protocol messages are idealised using the following rules:

1. Any nonsense is removed.
2. An atomic message is considered a response if it acts as a challenge as well as a response in a line.
3. Combine challenges separated by commas using operator 'I'.
4. Combine responses separated by commas using operator 'I' to form a combined response.
5. Combine a challenge and its response using 'R' into '*response* **R** *replied challenge*'.
6. Combine a message and its corresponding timestamp using 'R' into '*message* **R** *timestamp*'.

Finally, the following inference rules are also used:

1. *Authentication Rule*: If K is a shared secret key between P and Q, and P used K to decrypt a received message M, then P can believe that Q sent M. The rule is given as:

$$\frac{P \models P \overset{K}{\leftrightarrow} Q \wedge P \overset{K}{\triangleleft} M}{P \models Q \overset{K}{\mid\!\sim} M} .$$ (A.1)

2. *Confidentiality Rule*: If K is a shared secret key between P and Q and P encrypted M with K and sent it without sharing it with anyone else, then P can believe that M is only available to P and Q. The rule can be represented as

$$\frac{P \models P \overset{K}{\leftrightarrow} Q \wedge P \models S^{c} \triangleleft \| M \wedge P \overset{K}{\mid\!\sim} M}{P \models (S \cup \{Q\})^{c} \triangleleft \| M} .$$ (A.2)

3. *Super-Principal Rule*: P believes what Q believes if P believes Q is a trusted server. The rule is given as

$$\frac{P \models Q \models X \wedge P \models sup(Q)}{P \models X} .$$ (A.3)

This rule can be interpreted as P can trust Q about X. This means that an IoT device can be the super-principal for a nonce it generated. For example, in Protocol 1 IoT device ID_A generates N_A. Therefore, we can consider ID_A as the super principal in terms of N_A. This fact has been used in the tableaux of Figures 4.7 and 4.8.

4. *The Fresh Rule*: P can believe N is fresh if P believes M is fresh and P has received N and M together in a message. The rule is given as

$$\frac{P \models \#(M) \wedge P \triangleleft N \mathbf{R} M}{P \models \#(N)} .$$ (A.4)

5. *The Good-Key Rule*: There are two variations to this rule: (i) if P believes K is only available to P and Q, and P knows that K is fresh, then P can believe K is a good key between P and Q

$$\frac{P \models \{P,Q\}^{c} \triangleleft \| K \wedge P \models \#(K)}{P \models P \overset{K}{\leftrightarrow} Q}$$ (A.5)

and (ii) if P believes K is only available to P, Q and R and no one else, and P trusts R, and P knows K is fresh, then P can believe K is a good key between P and Q. The rule is given as

$$\frac{P\models\{P,Q,R\}^c\triangleleft\|K \wedge P\models sup(R) \wedge P\models\#(K)}{P\models P\overset{K}{\leftrightarrow}Q} . \tag{A.6}$$

6. *Intuitive Rule*: P has seen message M if it can decrypt message M using K. The rule is given as

$$\frac{P\overset{K}{\triangleleft}M}{P\triangleleft M} . \tag{A.7}$$

7. *Derived Rule*: This rule is obtained by combining the belief axiom

$$P \models (X \wedge Y) \text{ if and only if } P \models X \wedge P \models Y \tag{A.8}$$

with the confidentiality rule. The rule can be represented as

$$\frac{P\models Q\models P\overset{K}{\leftrightarrow}Q \wedge P\models Q\models S^c\triangleleft\|M \wedge P\models Q\overset{K}{\vdash}M}{P\models Q\models (S\cup\{P\})^c\triangleleft\|M} . \tag{A.9}$$

REFERENCES

1. Stankovic, J. A. 'Research directions for the Internet of Things.' *IEEE Internet of Things Journal* 1, no. 1 (February 2014): 3–9. ISSN: 2327-4662. doi:10.1109/JIOT.2014.2312291.
2. Verma, S., Y. Kawamoto, Z. M. Fadlullah, H. Nishiyama, and N. Kato. 'A survey on network methodologies for real-time analytics of massive IoT data and open research issues.' *IEEE Communications Surveys Tutorials* 19, no. 3 (third quarter 2017): 1457–1477. doi:10.1109/COMST.2017.2694469.
3. Anderson, R., and S. Fuloria. 'Who controls the off switch?' In *2010 First IEEE International Conference on Smart Grid Communications*, 96–101. October 2010. doi:10. 1109/SMARTGRID.2010.5622026.
4. Javaid, U., M. N. Aman, and B. Sikdar. 'BlockPro: Blockchain based data provenance and integrity for secure IoT environments.' In *Proceedings of the 1st Workshop on Blockchain-enabled Networked Sensor Systems*, 13–18. BlockSys'18. Shenzhen, China: ACM, 2018. ISBN: 978-1-4503-6050-0. doi:10.1145/3282278.3282281.
5. Suo, H., J. Wan, C. Zou, and J. Liu. 'Security in the Internet of Things: A review.' In *2012 International Conference on Computer Science and Electronics Engineering*, Vol. 3, pp. 648–651. March 2012. doi:10.1109/ICCSEE.2012.373.

6. Xu, T., J. B. Wendt, and M. Potkonjak. 'Security of IoT systems: Design challenges and opportunities.' In *2014 IEEE/ACM International Conference on Computer-Aided Design (ICCAD)*, 417–423. November 2014. doi:10.1109/ICCAD.2014.7001385.

7. Aman, M. N., K. C. Chua, and B. Sikdar. 'Mutual authentication in IoT systems using physical unclonable functions.' *IEEE Internet of Things Journal* 4, no. 5 (October 2017): 1327–1340. ISSN: 2327-4662. doi:10.1109/JIOT.2017.2703088.

8. Wood, D. D. Ethereum: A secure decentralised generalised transaction ledger, 2014. https://github.com/ethereum/yellowpaper

9. Karame, G. O., E. Androulaki, and S. Capkun. 'Double-spending fast payments in bitcoin.' In *Proceedings of the 2012 ACM Conference on Computer and Communications Security*, 906–917. CCS'12. Raleigh, North Carolina: ACM, 2012. ISBN: 978-1-4503-1651-4. doi:10.1145/2382196.2382292.

10. Nakamoto, S. 'Bitcoin: A peer-to-peer electronic cash system,' March 2009.

11. Beck, R., J. S. Czepluch, N. Lollike, and S. Malone. 'Blockchain the gateway to trust-free cryptographic transactions' [in English]. In *Twenty- Fourth European Conference on Information Systems (ECIS), Istanbul, Turkey, 2016*, 1–14. Springer Publishing Company, Istanbul, 2016. https://www.tib.eu/en/search/id/TIBKAT%3A875848443/24th-European-Conference-on-Information-Systems/

12. Buterin, V. *Ethereum: A Next-Generation Smart Contract and Decentralized Application Platform*. https://github.com/ethereum/wiki/wiki/White-Paper. Accessed: August 22, 2016, 2014.

13. Croman, K., C. Decker, I. Eyal, A. E. Gencer, A. Juels, A. Kosba, A. Miller et al. 'On scaling decentralized blockchains.' *Financial Cryptography and Data Security* 9604 (February 2016): 106–125. doi:10.1007/978-3-662-53357-4_8.

14. Aman, M. N., M. H. Basheer, and B. Sikdar. 'Two-factor authentication for IoT with location information.' *IEEE Internet of Things Journal* 6, no. 2 (April 2019): 3335–3351. ISSN: 2327-4662. doi:10.1109/JIOT.2018.2882610.

15. Aman, M. N., K. C. Chua, and B. Sikdar. 'Physical unclonable functions for IoT security.' In *Proceedings of the 2nd ACM International Workshop on IoT Privacy, Trust, and Security*, 10–13. IoTPTS'16. Xi'an, China: ACM, 2016. ISBN: 978-1-4503-4283-4. doi:10.1145/2899007.2899013.

16. Tosh, D., S. Shetty, X. Liang, C. Kamhoua, and L. L. Njilla. 'Data provenance in the cloud: A blockchain-based approach.' *IEEE Consumer Electronics Magazine* 8, no. 4 (July 2019): 38–44. ISSN: 2162-2248. doi:10.1109/MCE.2019.2892222.

17. Rosenfeld, K., E. Gavas, and R. Karri. 'Sensor physical unclonable fun tions' [in English (US)]. In *Proceedings of the 2010 IEEE International Symposium on Hardware-Oriented Security and Trust, HOST 2010*, 112–117. 2010. ISBN: 9781424478101. doi:10.1109/HST.2010.5513103.

18. Guilley, S., and R. Pacalet. 'SoCs security: A war against side-channels.' *Annals of Telecommun.*, 59, no. 7 (2004): 998–1009.

19. Kirkpatrick, M., S. Kerr, and E. Bertino. System on chip and method for cryptography using a physically unclonable function. Patent US 8750502 B2 (US), filed issued March 22, 2012.

20. Bohm, C., and M. Hofer. *Physical Unclonable Functions in Theory and Practice.* Springer-Verlag, New York, 2012. ISBN: 146145039X, 9781461450399. https://www.springer.com/gp/book/9781461450399.
21. Mao, W., and C. Boyd. 'Towards formal analysis of security protocols.' In *[1993] Proceedings Computer Security Foundations Workshop VI,* 147–158. June 1993. doi:10.1109/CSFW. 1993.246631.
22. Abadi, and Needham. 'A logic of authentication.' *ACM Transactions on Computer Systems,* no. 8 (February 1990): 18–36. doi/10.1145/77648.77649.
23. Popov, S. 'The tangle.' 2015. https://www.iota.org/research/academic-papers.

Chapter 5

On blockchain and its complementarity with fog computing to secure management systems

Chaima Khalfaoui, Samiha Ayed, and Moez Esseghir

CONTENTS

5.1 INTRODUCTION

The Internet of things (IoT) allows billions of limited power devices to interconnect every day in the aim of enhancing data exchange and automating daily services. However, to manage these huge amounts of data, many security and performance challenges must be addressed. Such a challenge requires a solution with efficient decision making, short delays, and data protection. We propose an architecture based on blockchain technology with the use of fog computing to increase data security and user privacy with the shortest delays. The use of the fog computing paradigm allows the support of the high computing demand of IoT applications. Therefore, the security of this infrastructure is still an open issue mainly because of the lack of mechanisms able to secure information in a such untrustworthy environment. In this context, we propose the collaboration of blockchains with the fogs to fill the security gap and ensure effective services. There are recent works [1,2] that studied the use of fogs or blockchains separately to improve computing performances or security. However, privacy and security issues were not addressed for works associating both fogs and blockchains paradigms simultaneously. In this work, we propose the collaboration between blockchain technology and fog computing through a three-layer architecture aiming to exchange and manage data securely and effectively.

5.2 IoT CONTEXT

The IoT is 'the network of physical objects, devices, vehicles, buildings and other items which are embedded with electronics, software, sensors, and network connectivity, permitting these objects to gather and interchange data' [3–5]. There is no doubt that the IoT network is playing a major role in the everyday life of users through the development of a huge number of applications in different fields such as transportation, health care, and smart environments or assisted systems today. Thanks to IoT, it becomes possible to monitor your home and enterprise remotely and even to keep an ambient climate in your home through temperature sensors exchanging data with remote services. Real-time data sensing functions and automatic data collection and transfer are providing innumerable benefits to improve the quality of services.

Despite the benefits mentioned, IoT scenarios require specific architecture and measurements. More precisely, network and security issues have to be addressed. Let's start by identifying the main attributes of IoT devices:

- *Lack of security*: IoT devices are vulnerable to security threats because they have low energy and computation capabilities enable to support robust security algorithms.
- *Heterogeneity*: IoT devices deployed in a system are heterogeneous and based on multiple networks most of the time. Therefore, the key requirement is in deploying an architecture that supports heterogeneous direct connectivity.
- *Huge continuous increase in number of devices*: Is it predicted to be 18 billion devices by 2022 [6]. A scalable platform needs to be designed to meet the requirements.

In the health care domain, most of the current systems manipulating sensitive data [7,8] are still suffering from the lack of efficient security mechanism to protect patient medical information or the low performance prevents IoT devices to work with short delays. Besides the challenge of storing and sharing, to effectively detect and fix struggles in the current systems, medical data characteristics have to be investigated. The main characteristics are as follows:

- *Data integrity*: Clinical data need to be trusted, reliable, and free of contradiction; data precision and accurateness are two essential features for doctors to provide effective treatments. In addition, data sources have to be determined (authenticated).
- N *Data privacy and confidentiality*: The protection of patient privacy, identity, and medical records are primary requirements. The protection of sensitive data has to be considered to provide reliable services.

However, the majority of medical organisations are still having security issues. On the one hand, destroying the security of a medical system or network could cause disastrous consequences. On the other hand, the patient's privacy information exists at all stages of data collection, data transmission, cloud storage, and data republication [8].

As aforementioned, several challenges may hinder the successful deployment of IoT components including the lack of security, the limited capabilities, and scalability. However, traditional available mechanisms in relation with data management and security do not meet current IoT requirements. To achieve efficient access and use of the smart objects, several related works about IoT technology considered cloud computing technology. Cloud computing is an Internet-based computing paradigm that provides ubiquitous and on-demand access to a shared pool of configurable resources (e.g., processors, storage, services, and applications) to other computers or devices. The cloud computing paradigm is able to handle huge amounts of data from IoT clusters, that is, the transfer of enormous data to and from cloud computers [9]. Tyagi et al. in [10] conceived an IoT-cloud health care system aiming to improve the communication and collaboration between the different medical entities (doctors, patients, medical staff, etc.) and to provide better health care services to the patients. However, results showed that such a system still needs a complete model to tackle security, privacy, and trust issues. Another research about the integration of cloud computing and IoT [11] analysed the possible challenges derived from such integration and resume that:

1. Data ownership and privacy must be addressed.
2. Energy management is a major issue.
3. New techniques such as big data are required to optimise and processing large volumes of heterogeneous data quickly.
4. Security requirements are a major issue on both sides because it is difficult to implements security mechanism on computation constrained devices and cloud quality of service (QOS) and management algorithm have to be enhanced.

To conclude, cloud computing is facing several problems in ensuring secure and efficient services, especially the real-time applications because of specific IoT requirements such as resource and energy limitation and the cloud servers' distant geolocation.

5.3 FOG COMPUTING

As detailed before, cloud computing has several benefits such as managing, storing, and treating data. By providing cheap and rapid scaling performances, it spares enterprises unnecessary details such as the number of required resources or storage limits. However, it is another paradigm that ensures low latency.

5.3.1 Fog computing strengths and weaknesses

Fog computing has emerged as an extension of the services that offers cloud computing to the edge of the network. This paradigm is proposed to support heterogeneous, dense, and distributed data used in latency-sensitive and real-time applications. It is an advantageous approach dealing with the IoT environment because of the following technology concept criteria:

- *Offering low-latency*: Placing processing on the network edge, thus in a closer physical location to the user, allows speeding up computation and data transmission, which leads to shorten response time and reduces latency.
- *Saving bandwidth use*: Only a portion of the data is transmitted to the cloud when necessary, and the rest of it is analysed and processed locally.
- *Disposing of high-processing capacities*: Fog networks consist of a large number of nodes communicating and interoperating to offer an efficient distributed monitoring and computing services.

In addition, high network and hardware scalability are challenging in the cloud layer for monetary reasons and in IoT devices because of their limited computational power and capacities. Conversely, fog nodes are more likely to satisfy large-scale applications. In [12], the authors declare that each fog node is only needed to satisfy a small part of this requirement. As the number of IoT devices is growing, the increased network bandwidth and computing capability requirement can be easily satisfied by adding fog nodes in fog computing.

However, several reviews studied the impact of fog computing deployment in different domains. In 'Fog Computing in Healthcare – A Review and Discussion' the authors showed how health care applications can benefit from these technology principles. With concrete examples, they explained how end devices are unable to handle computation tasks such as home treatment or data monitoring and control because they are power-energy limited. On the other hand, cloud computation is not suitable for such tasks because of dependability and centralisation concerns. Fog computing, with its flexibility to add computation as part of a network infrastructure, appears as a suitable concept to meet the requirements of health care. Fog computing tasks can filter data to help preserve privacy or reduce load on the network. The locus of execution can be adjusted to the current deployment scenario, regulations, and other requirements [13].

One of the main security problems in fog is authentication at each level of the devices (servers, gateways, etc.). Each node has its own address and an attacker could tamper or spoof that address. As explained in [14,15], man-in-the-middle (MIM) attacks have the potential to become a typical attack in fog. Some authentication protocols such as public key infrastructure (PKI)-based approaches and intrusion detection techniques can be considered solutions. In addition, privacy issues and the risk of privacy leakage of user information (e.g., personal data, location) have to be studied. Fog computing can adopt privacy-preserving algorithms using, for example, cryptography to protect data because they have sufficient computing power.

Whether it is authentication, access control, or privacy issues, the major challenge about securing fog is how to secure the source (devices) and paths before reaching the fog nodes. As mentioned, IoT devices are resource constrained; thus, it is difficult to devise algorithms for them. As far as we know, there are no infrastructure able to tackle all the mentioned security issues without altering the fog calculation capabilities.

5.4 BLOCKCHAINS

Blockchain is an emerging innovative technology of communication and cyber security. In 2008, its germinal state, Blockchain was proposed as a peer-to-peer cryptocurrency named 'Bitcoin'. With the development of its concept and components, blockchain can be defined as a decentralised distributed public ledger able to ensure data exchange and management in a high secure and verifiable way without the need of a third party. Blockchains can store anything of value starting with row data to financial transactions. The technology has successfully replaced economic transaction systems in various organisations and has the potential to revamp heterogeneous business models in different industries. It promises a secure distributed framework to facilitate sharing, exchanging, and integrating information across

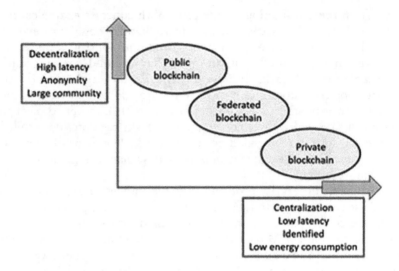

Figure 5.1 Blockchain classification schema.

all users and third parties [16]. There are three main types of blockchains depending on who has access to the ledger as shown in Figure 5.1:

- *Public blockchain*: The most commonly known because of the rise of Bitcoin and Ethereum. Public blockchains are fully decentralised, openly available, and free to download and use so anyone can be part of it. For example: Bitcoin[1] cryptocurrency, Litecoin,[2] Ethereum.[3]
- *Private blockchain*: Are a private property of an individual or an organisation. All the rights and permissions are kept by one entity such as a company who wants to keep its data private, for example, Monax[4] for business and digital contracts.
- *Hybrid or consortium or federated blockchain*: This blockchain is characterised as partially decentralised because participating nodes are preselected and decisions are defined by an authorised set of nodes. They are mostly used in the banking sector, for example, R3[5] for business and B3i[6] for insurance.

In practice, blockchain is actually a database because it is a digital chain that stores timestamped and structured data called 'blocks'. Once the

---

[1] https://bitcoin.org/fr/
[2] https://litecoin.org/
[3] https://www.ethereum.org/
[4] https://monax.io/
[5] https://www.r3.com/
[6] https://b3i.tech/home.html

data is recorded in the blockchain, it becomes indestructible. It is difficult to tamper or delete recorded data because of the blockchain structure itself. Blocks are linked to each other, forming a continuous chain. A database likewise stores information in data structures called 'tables'. However, although a blockchain is a database, a database is not a blockchain. They are not interchangeable in a sense that though they both store information, they differ in design.[7]

5.4.1 What is blockchain structure?

The blockchain is an ordered back-linked list of blocks carrying sets of data called 'transactions' created and submitted by entities called 'miner node's. For example, in the field of clinical data management, a transaction may be the addition of a new patient to the database, the registration of the next appointment, or the recording of a new clinical prescription. Each block is identified by a hash of its header and contains other information that are of paramount importance for other peers to easily verify its correctness [17]. In addition, each block references a previous block, also known as the 'parent block', in the 'previous block hash' field. Thus, a previous block hash represents a pointer to the previous block.

As shown in Figure 5.2, a standard transaction is a structure that contains verifiable value and metadata: timestamp to indicate creation time, one or more inputs as references to the sender addresses, one or more output

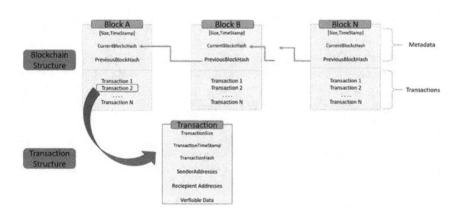

Figure 5.2 Blockchain structure.

---

[7] https://hackernoon.com/databases-and-blockchains

as references to recipient addresses, the size of the transaction, and the signature of the sender to prove that sender is the creator of the transaction.

5.4.2 How does blockchain work?

Miners are the main users of the blockchain; they access to their accounts in the chain using their unique addresses. Each miner has a shared copy of the register and owns the right to add blocks and, therefore, enrich the distributed database according to the blockchain rules (mining consensus algorithm). Consensus is fixed and enriched depending on the blockchain nature and its users' requirements. A typical algorithm can be defined as follows:

1. New transactions created by miners are broadcast to all the network nodes.
2. Each connected node is continuously listening to the network to collect new transactions into a block.
3. In each round, according to the chosen consensus protocol defined by the system, one node gets to broadcast its block.
4. Other nodes accept new blocks only if the transactions are valid (e.g., unspent, legitimate, with valid signatures, etc.).
5. Nodes validate and express their acceptance by including the current block hash in the next block they create.

As mentioned in step 3, blocks are not broadcast randomly; miners follow the consensus protocol usually hard coded in the system. In practice, proof-of-work (PoW) is the most popular and stable algorithm. In the POW algorithm, miners calculate a hash value constantly changing depending on the network environment (e.g., number of nodes, required time to add a block). The algorithm requires that the hash value must be equal to or smaller than a certain given value fixed by the system.

Participants are continuously seeking the value using random values called 'nonce' until the target is reached. Once they find the required nonce, they can publish the result. The key to this protocol is that the search process must be moderately hard for the miners who build the blocks but easy for the miners who validate it to check.

However recently, several other consensus algorithms have been proposed such as proof-of-stake (POS), proof-of-authentication, and proof-of-concept [18–20].

The preceding algorithm is simplified; in real-life blockchain use, transactions are not directly preceded, and the wait in a pool is called a' transaction pool'. The pool of unconfirmed transactions at each node, thus, consists of legitimate and inconsistent transactions. The distributed transactions then wait in a queue, or transaction pool, until they are added to the blockchain by a mining node [21,22].

5.4.3 What are blockchain key features?

Blockchains have the following properties: immutable (consensus mechanism), traceable (shared ledger), decentralised (peer-to-peer network), and pseudonymous (pseudo-identity). In [23,24], the authors detail each feature and explain the mechanisms behind its development.

- *Decentralised*: There is no central authority to determine rules. Instead, all authenticated nodes have access to the ledger according to their permissions. The information is transparent and saved in blocks in a chronological order as a result of the consensus mechanism.
- *Immutable*: The blockchain system can be considered trusted because the content of the ledger is agreed on by all the parties and an updated copy is diffused between the nodes. This mechanism makes it extremely difficult to be tampered with or manipulated by malicious users unless they control the majority of the nodes.
- *Traceable*: In blockchain, it is possible to trace the history of all the transactions. This feature is feasible thanks to the immutable characteristic described. Because users cannot modify the content of the chain, all the history is available.
- *Pseudonymous*: Blockchain offers relative user anonymity using public key as unique identifiers and keeping the real user's identity hidden. In addition, the affected pseudonym does not contain any indication or identifiable information about its user. Hence, the mechanism provides relative user privacy.

5.4.4 What are blockchain performance and security weaknesses?

Unfortunately, as any type of technology, blockchain still suffers from important weaknesses that hinder its adoption in a real context and especially in an IoT environment. One of the major weak points is energy consumption.

The mining process consumes a high volume of energy because the computationally intensive POW algorithm. Some works try to eliminate, replace, or improve the POW mechanism. To our knowledge, these attempts partially solve the problem because they are always at the cost of relaxing other features and, most usually, security properties (e.g., anonymity, immutability, etc.). In contrast, the resource-constrained architecture of IoT makes it difficult to deploy the technology. Adapted mechanisms have to be lightweight to work efficiently within these constraints.

The second key drawback is high latency for transaction confirmations, especially because we are in an IoT environment, and most of the recent applications are aiming to provide real-time services (e.g., transportation, IIoT, etc.): Compared to traditional online credit card transaction, which usually takes two or three days to confirm the transaction, a Bitcoin transaction only has to use about one hour to verify; it is much better than the usual, but it is still not good enough for what is needed [25].

Several recent works discuss the possibility of combining blockchains with other technologies to address performance issues. Seitz et al. [1] exploit the cloud and fog paradigms to design a blockchain-based application. In their prototype, they deployed a blockchain in the cloud layer for a tamper-proof storage and used one fog as a proxy between the cloud and the devices. Nevertheless, several limitations have been identified, among which the most significant is the high confirmation times even though they used a proof-of-authority instead of a POW. Dorri et al. [26] introduced a lightweight block-chain eliminating the mining protocol to decrease processing overheads.

We also need to point out that blockchains present various security gaps. By default, they do not include access control policies or authentication ones. Furthermore, the protection of the private key is the responsibility of the owner. If it happens that the key is stolen, the thief may gain access to the chain and start a sniffing attack of even a distributed denial-of-service (DDoS) attack sending a huge number of useless but valid transactions aimed at weighing down the network.

However, blockchains are continually evolving, and smart contracts are one of the most revolutionary progresses.

5.4.5 What is a smart contract?

In [27] smart contracts are defined as computer protocols that are designed to automatically facilitate, verify, and enforce the negotiation and imple-mentation of digital contracts without central authorities. Cong et al. [28] see it as an automated and self-enforcing digital program relying on tamper-proof consensus on contingent outcomes and financing through initial coin offerings. Both definitions amount to saying that smart contracts are tiny programs developed and deployed in the blockchain networks with the only aim to allow the execution of trusted agreements without the need of inter-mediaries. Several fields are testing blockchains-based smart contracts. To illustrate my point, the health care domain, [24,29,30] deployed smart con-tracts to help data management, exchange, and monitoring. The Ethereum platform is now a leader in smart contracts. Solidity[8] is the programming language used to implement them.

5.5 BLOCKCHAIN AND FOG COMPLEMENTARITY

Both fog and blockchain provide data exchange and computation services to end users. In addition, they represent mostly the same limitations. However, each technology can be distinguished by its unique features that matched together can be described as complementary as shown in Figure 5.3.

---

[8] https://solidity.readthedocs.io

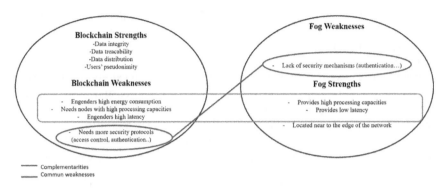

Figure 5.3 Blockchain-fog complementarity and specificities.

Blockchain provides many opportunities by offering its users a decentralised exchange environment ensuring integrity and traceability with no extra fees. Nevertheless, based on the latest developed frameworks and proposed solutions detailed previously, we noticed that performance constraints such as long delays and the lack of security are preventing or at least slow downing the technology from large-scale deployment in applications with real-word requirements.

In blockchains, the consensus defines the latency at which the blocks are created. Particularly, system latency depends mostly on the information verification and propagation mechanisms. One solution would be to lighten the mechanism. Thus, reducing consensus algorithm directly affects the security of the system. Because the mechanism process assumes that the more powerful the miner is (processing power) the faster the process runs, it would be wiser to focus on providing a solution able to handle this requirement rather than replacing the consensus.

Taking into consideration that IoT devices have been designed as low-power technology with no processing capacities, it is certain that integrating them directly as nodes will not allow an efficient process execution. In fogs, however, the decentralised computing infrastructure is distinguished by high-processing capacities, and the scalability aims to improve the response time and minimise the network latency. Unfortunately, although introduced as one of the most suitable platforms dedicated to facilitate data collecting, computing, and storage from IoT devices, fogs still suffer from several limitations. Basically, the primary model of fogs lacks security. This can be seen through the lack of mechanisms for identity management, access management or even for the protection of private data.

In this work, we analysed the performance and security features and issues in both technologies to show the significant compatibility between both of them.

Additionally, we analysed works related to our proposal and using the same two key technologies in their architectures. In [31], the authors adopted cloud computing and blockchains solutions with fogs to replace databases rule validations by the consensus and smart contracts. However, privacy and security issues were not addressed simultaneously. As in [1], the authors studied the blockchain integration into a cloud structure to ensure a better traceability and defend against security attacks such as MIM. However, the results showed long delays because of the time of implementation of the consensus. In [14], the authors adopted cloud computing and blockchain solutions to replace databases rule validations by the consensus between validators and smart contracts. In [15] they studied the use of fogs to resolve the same issues. However, privacy and security issues were not addressed simultaneously.

We conclude that blockchains have the potential to address some important issues in fogs by providing additional protection mechanisms such as data integrity. Although, to ensure an efficient and secure data management, there are other security challenges to be tackled not only concerning the fog layer but also derived from the nature of IoT devices such of the lack of solid authentication algorithms.

5.6 PROPOSITION

In this work, we propose to tackle some of the issues by adopting a three-level architecture. We provide an overview of the architecture design in Figure 5.4.

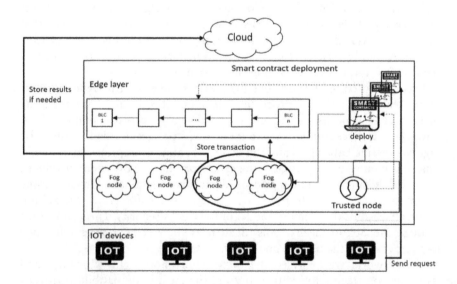

Figure 5.4 Proposed architecture.

Our system is composed of: (i) IoT devices layer, (ii) the cloud computing layer considered safe in our architecture, and (iii) the network layer based on fog computing and blockchains.

The goal of this proposal is to take advantage of the performance and security characteristics of the components to provide an efficient and secure data management system that can handle the exponential volume of data generated from IoT devices and face the potential attacks.

To this end, we identify the three main components of particular importance in the network layer.

1. The first component is the blockchain and its three smart contracts. The blockchain is responsible for data orchestration, access, and storage using smart contracts. In this architecture, we choose to use a public blockchain to guarantee full decentralisation.
2. Second, a trusted entity is the only component in the network with rights to create and deploy the smart contracts.
3. The trusted entity is not a full node in the blockchain, the main role is to deploy and monitor the smart contracts. Blockchain nodes have only the right to execute already added smart contracts.
4. Third, fog nodes are considered the blockchain nodes. They execute services, execute service mechanisms, and save results in the blocks and then send results back to the user or the cloud if required. This choice is one of the efficiency keys in our architecture. We took advantage of the high processing power to consider them the blockchain nodes.

The distribution of our architecture allows blockchains to benefit from fogs' computational power considering them as miner nodes to minimise the execution time of the consensus mechanism. On the other hand, thanks to the consensus and cryptography protocols of the blockchain network aiming to ensure data integrity, fog nodes can exchange data securely without fear of tampering attacks. In addition, as part of the network, fogs will have to authenticate before accessing the data, respecting the pseudonymous feature imposed by the chain.

For access control, confidentiality, and data balancing issues, we decided to not include external components but take advantage of the new powerful logical component of the blockchains 'smart contracts'.

As mentioned and as described in several recent works [32,33], smart contracts are self-executing small programs that have the advantage to be included in the blockchain to bend agreements between different parties. Hence, the security requirements can be offset by the use of policies written in the smart contracts because they are described by Vitalik Buterin as 'cryptographic' boxes 'that contain value and only unlock it if certain conditions are met'.

To satisfy the mentioned security requirements and enhance the performance of the system, we adopt three smart contracts:

- *The first smart contract aims to define a data access control policy*: To access the data in the ledger, once authenticated, fog nodes and new devices have to register under an access permission list and get permissions according to their rights.
- *The second smart contract is responsible for data confidentiality in the blockchain network*: Using a cryptographic PKI integrated in the contract, data arrive encrypted to the network and the smart contract decrypts it to execute users' transactions.
- *The third smart contract concerns the performance of the system*: It defines a data orchestrating policy aiming to optimise transaction assignment. For a better understanding of the process, we propose the following datagram in Figure 5.5. The datagram explains the execution steps of a service in the system starting with sending a request to storing in the cloud. Orchestration and optimisation algorithms are needed to be set to support the classical process of transaction assignments in blockchains. Instead of broadcasting the tasks to all the nodes and letting them choose the transaction they want, the smart contract will determine a group of nodes that will operate depending on the nature of the service to execute and the capacities of the fogs.

In this way, we achieve two goals at the same time: first, a fair sharing of the tasks, and second, a better monitoring of the fogs access rights because they no long have the right to access all the transactions.

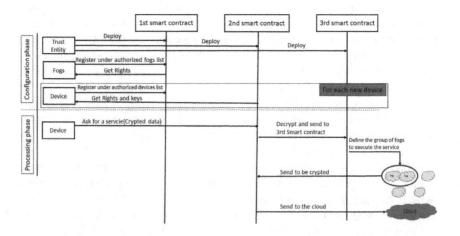

Figure 5.5 Service execution process.

To summarise, a number of contributions study the use of fogs and blockchains in the same architecture, but they do not consider their collaboration in a single layer. In this work, our assumption is fog computing and blockchains are complementary in terms of performance and security and that several security issues that are not considered in the blockchain structure may be tackled using smart contracts.

In this work we proposed an architecture that aims to ensure, on the one hand, the protection and security such as integrity, authentication, and confidentially of the data and enhance the system performance such as delays, on the other hand.

- Data in the cloud is encrypted, therefore, even if an attacker compromises the content of the cloud, it would not cause data leakage. Furthermore, data access is restricted to authorised and trusted entities and decryption keys are stored in the smart contract of the blockchain, which make it difficult to steal.
- *On the network*: We mainly use blockchains to enhance data protection and security. Transaction process is protected by blockchain PKI so that only authorised entities may communicate and read the data, decreasing the risk of MIM and spoofing attacks. Updated copies of the ledger increase system immutability and viability and defend against single point of failure attack. If a node is compromised or down, the system is able to continue working correctly because of redundancy.
- *Between layers*: Data in transition is encrypted with PKI keys and keys and transmitted in a secure channel. Each device is represented with a unique identity in the blockchain and may hinder attackers from leading a linking attack and increase user privacy.

However, blockchains are vulnerable to other type of attacks that are not considered in this work. These attacks are related to the blockchain basic structure. The most famous ones are: the race attack, 51% attack, and withholding block attack. The authors of [34,35] analyse in detail the selfish mining and the 51% attack and their effect on the system.

5.7 CONCLUSION

This chapter presents a decentralised approach for localisation of sensors in indoor wireless networks. At first, a decentralised architecture is constructed, partitioning the target area into sectors, each having its own calculator. A local localisation algorithm is then run by each calculator using the belief functions theory. Afterwards, the calculator's estimations are fused, resulting in an observation model. The proposed algorithm then uses the mobility of sensors to assign another type of evidence. Levels of confidence are attributed to zones by fusing evidence from both observation and mobility models. Experiments are conducted to evaluate the performance of the decentralised approach and compare it against a centralised one. In addition, the accuracy, complexity, robustness, and the influence of detected access points on the localisation algorithm are studied in comparison to well-known techniques in the domain. Future work will focus on kernel-based models for cases where parametric distributions fail to represent the received signal strength indicator (RSSI). In addition, advanced mobility models such as hidden Markov models will be studied to enhance the overall localisation algorithm.

REFERENCES

1. A. Seitz, D. Henze, D. Miehle, B. Bruegge, J. Nickles, and M. Sauer, 'Fog computing as enabler for blockchain-based IIoT app marketplaces-a case study,' in *2018 Fifth International Conference on Internet of Things: Systems, Management and Security*, pp. 182–188, IEEE, 2018.
2. A. Stanciu, 'Blockchain based distributed control system for edge computing,' in *2017 21st International Conference on Control Systems and Computer Science (CSCS)*, pp. 667–671, IEEE, 2017.
3. A. Munir, P. Kansakar, and S. U. Khan, 'IFCIoT: Integrated Fog Cloud IoT: A novel architectural paradigm for the future Internet of Things,' *IEEE Consumer Electronics Magazine*, vol. 6, no. 3, pp. 74–82, 2017. doi:10.1109/MCE.2017.2684981.
4. L. Atzori, A. Iera, and G. Morabito, 'The Internet of Things: A survey,' *Computer Networks*, vol. 54, no. 15, pp. 2787–2805, 2010.
5. M. Samaniego and R. Deters, 'Blockchain as a service for IoT,' in *2016 IEEE International Conference on Internet of Things (iThings) and IEEE Green Computing and Communications (GreenCom) and IEEE Cyber, Physical and Social Computing (CPSCom) and IEEE Smart Data (SmartData)*, pp. 433–436, IEEE, 2016.
6. O. Novo, 'Blockchain meets IoT: An architecture for scalable access management in IoT,' *IEEE Internet of Things Journal*, vol. 5, no. 2, pp. 1184–1195, 2018.
7. P. Zhang, J. White, D. C. Schmidt, G. Lenz, and S. T. Rosenbloom, 'Fhirchain: Applying blockchain to securely and scalably share clinical data,' *Computational and Structural Biotechnology Journal*, vol. 16, pp. 267–278, 2018.
8. W. Sun, Z. Cai, Y. Li, F. Liu, S. Fang, and G. Wang, 'Security and privacy in the medical Internet of Things: A review,' *Security and Communication Networks*, vol. 2018, 9 pages, 2018 doi:10.1155/2018/5978636.
9. C. Stergiou, K. E. Psannis, B.-G. Kim, and B. Gupta, 'Secure integration of IoT and cloud computing,' *Future Generation Computer Systems*, vol. 78, pp. 964–975, 2018.
10. S. Tyagi, A. Agarwal, and P. Maheshwari, 'A conceptual framework for IoT-based healthcare system using cloud computing,' in *2016 6th International Conference-Cloud System and Big Data Engineering (Confluence)*, pp. 503–507, IEEE, 2016.

11. A. Botta, W. De Donato, V. Persico, and A. Pescapé, 'On the integration of cloud computing and Internet of Things,' in *2014 International Conference on Future Internet of Things and Cloud*, pp. 23–30, IEEE, 2014.
12. Y. Guan, J. Shao, G. Wei, and M. Xie, 'Data security and privacy in fog computing,' *IEEE Network*, vol. 32, no. 5, pp. 106–111, 2018.
13. F. A. Kraemer, A. E. Braten, N. Tamkittikhun, and D. Palma, 'Fog computing in healthcare–A review and discussion,' *IEEE Access*, vol. 5, pp. 9206–9222, 2017.
14. I. Stojmenovic and S. Wen, 'The fog computing paradigm: Scenarios and security issues,' in *2014 Federated Conference on Computer Science and Information Systems*, pp. 1–8, IEEE, 2014.
15. I. Stojmenovic, S. Wen, X. Huang, and H. Luan, 'An overview of fog computing and its security issues,' *Concurrency and Computation: Practice and Experience*, vol. 28, no. 10, pp. 2991–3005, 2016.
16. D. Puthal, N. Malik, S. P. Mohanty, E. Kougianos, and C. Yang, 'The blockchain as a decentralized security framework [future directions],' *IEEE Consumer Electronics Magazine*, vol. 7, no. 2, pp. 18–21, 2018.
17. V. Daza, R. Di Pietro, I. Klimek, and M. Signorini, 'Connect: Contextual name discovery for blockchain-based services in the IoT,' in *2017 IEEE International Conference on Communications (ICC)*, pp. 1–6, IEEE, 2017.
18. A. Dorri, M. Steger, S. S. Kanhere, and R. Jurdak, 'Blockchain: A distributed solution to automotive security and privacy,' *IEEE Communications Magazine*, vol. 55, no. 12, pp. 119–125, 2017.
19. Z. Zheng, S. Xie, H.-N. Dai, X. Chen, and H. Wang, 'Blockchain challenges and opportunities: A survey,' *International Journal of Web and Grid Services*, vol. 14, no. 4, pp. 352–375, 2018.
20. D. Puthal, S. P. Mohanty, P. Nanda, E. Kougianos, and G. Das, 'Proof-of-authentication for scalable blockchain in resource-constrained distributed systems,' in *2019 IEEE International Conference on Consumer Electronics (ICCE)*, pp. 1–5, IEEE, 2019.
21. D. Yaga, P. Mell, N. Roby, and K. Scarfone, 'Blockchain technology overview,' arXiv preprint arXiv:1906.11078 (2019). doi:10.6028/NIST.IR.8202. 2018.
22. E. O. Kiktenko, N. O. Pozhar, M. N. Anufriev, A. S. Trushechkin, R. R. Yunusov, Y. V. Kurochkin, A. Lvovsky, and A. Fedorov, 'Quantum-secured blockchain,' *Quantum Science and Technology*, vol. 3, no. 3, p. 035004, 2018.
23. R. Beck, J. Stenum Czepluch, N. Lollike, and S. Malone, 'Blockchain–the gateway to trust-free cryptographic transactions,' 2016. https://aisel.aisnet.org/ecis2016_rp/153/.
24. X. Li, P. Jiang, T. Chen, X. Luo, and Q. Wen, 'A survey on the security of blockchain systems,' *Future Generation Computer Systems*, vol. 107, pp. 841–853, 2017.
25. I.-C. Lin and T.-C. Liao, 'A survey of blockchain security issues and challenges.,' *IJ Network Security*, vol. 19, no. 5, pp. 653–659, 2017.
26. A. Dorri, S. S. Kanhere, and R. Jurdak, 'Towards an optimized blockchain for IoT,' in *Proceedings of the Second International Conference on Internet-of-Things Design and Implementation*, pp. 173–178, ACM, Pittsburgh, PA, 2017.

27. S. Wang, L. Ouyang, Y. Yuan, X. Ni, X. Han, and F.-Y. Wang, 'Blockchain-enabled smart contracts: Architecture, applications, and future trends,' *IEEE Transactions on Systems, Man, and Cybernetics: Systems*, vol. 49, pp. 2266–2277, 2019.
28. L. W. Cong and Z. He, 'Blockchain disruption and smart contracts,' *The Review of Financial Studies*, vol. 32, no. 5, pp. 1754–1797, 2019.
29. K. N. Griggs, O. Ossipova, C. P. Kohlios, A. N. Baccarini, E. A. Howson, and T. Hayajneh, 'Healthcare blockchain system using smart contracts for secure automated remote patient monitoring,' *Journal of Medical Systems*, vol. 42, no. 7, p. 130, 2018.
30. A. Azaria, A. Ekblaw, T. Vieira, and A. Lippman, 'Medrec: Using blockchain for medical data access and permission management,' in *2016 2nd International Conference on Open and Big Data (OBD)*, pp. 25–30, IEEE, 2016.
31. R. B. Uriarte and R. De Nicola, 'Blockchain-based decentralized cloud/fog solutions: Challenges, opportunities, and standards,' *IEEE Communications Standards Magazine*, vol. 2, no. 3, pp. 22–28, 2018.
32. H. Watanabe, S. Fujimura, A. Nakadaira, Y. Miyazaki, A. Akutsu, and J. Kishigami, 'Blockchain contract: Securing a blockchain applied to smart contracts,' in *2016 IEEE International Conference on Consumer Electronics (ICCE)*, pp. 467–468, IEEE, 2016.
33. S. Porru, A. Pinna, M. Marchesi, and R. Tonelli, 'Blockchain-oriented software engineering: Challenges and new directions,' in *2017 IEEE/ACM 39th International Conference on Software Engineering Companion (ICSE-C)*, pp. 169–171, IEEE, 2017.
34. I. Eyal and E. G. Sirer, 'Majority is not enough: Bitcoin mining is vulnerable,' *Communications of the ACM*, vol. 61, no. 7, pp. 95–102, 2018.
35. D. Efanov and P. Roschin, 'The all-pervasiveness of the blockchain technology,' *Procedia Computer Science*, vol. 123, pp. 116–121, 2018.

Chapter 6

Security and privacy issues of blockchain-enabled fog and edge computing

Imane Ameli, Nabil Benamar, and Abdelhakim Senhaji Hafid

CONTENTS

6.1 INTRODUCTION

Information technology is omnipresent and has become an integral part of our daily activities. All sorts of sophisticated devices and their applications are in rapid evolution. Those devices provide new opportunities to be practically leveraged in a broad spectrum of applications in different areas, such as home automation, health care, smart agriculture, supply chain and logistics, Smart cities, smart grid, connected cars, and wearables. These applications generate huge amounts of data that are usually transmitted to centralised cloud servers for processing and storing. According to Cisco Internet Business Solutions Group experts, the number of connected devices is expected to reach 500 billion by 2030 [1].

The current cloud computing frameworks, such as Amazon Web Services and GoogleApp Engine, can support processing-intensive applications. However, with the rate, distribution, and scale of extraordinary data generated by connected devices, processing all the data in the cloud would be impractical. Particularly, when it comes to applications with stringent delay and high quality of services (QoS) requirements, cloud computing cannot adequately address these requirements; for example, road safety and autonomous driving applications require latencies of less than 50 ms [2], while smart factories have even stricter requirements with latencies [3].

Moreover, sending all data to the cloud would cause an explosion of network traffic. Generally, it is not efficient because there is considerable redundancy in the data. It is even impossible given the huge amount of traffic data that can exceed the capacity of the network. To overcome the limitations of cloud computing and successfully support a wide range applications, a new computing paradigm has been proposed to move data processing to the proximity of devices and users from the core to the edge of the network.

In this context, fog has been introduced by Cisco as an extension of cloud computing [4]. ARM, Cisco, DELL, Intel, and Microsoft share the same vision about the future and the utility of fog computing. They established the OpenFog consortium which aims to speed up the development and deployment of fog computing infrastructure [5]. Fog computing and edge computing are both concerned with moving data processing close to the end devices; in this chapter, we use the term 'fog computing' even though there are differences between the two based on where the compute power is placed (see Section 6.2). With fog computing, processing data is performed closer to the edge, and only digests or exceptions are sent into the cloud for more processing or storage [6]. This provides the necessary support to process time-sensitive data where real-time decisions are needed. In the case of the cloud, the time taken to transfer the data is high; thus, the decision making can no longer be timely enough. Fog computing is a powerful alternative to the cloud to process all generated data at the fog nodes. It is a layer placed, between the cloud and

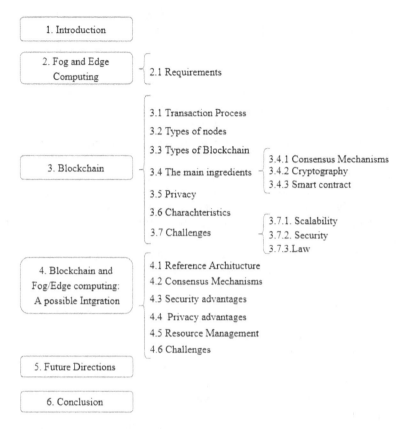

Figure 6.1 Diagram classification.

end devices, as a bridge that provides storage, computation, and networking services (Figure 6.1). This allows supporting a wide range of Internet of Things (IoT) applications such as video streaming and gaming, and so on by moving the processing to the proximity of users and processing data at the network edge or anywhere along the IoT-fog-cloud continuum that can best meet the requirements of applications. Indeed, fog computing inherits and extends the benefits of cloud computing [7]. Equally of importance are the privacy and security of data generated by IoT devices, which are critical factors to gain the confidence of application users. Indeed, vulnerabilities exist in *every* system, including fog computing. Hackers always try to exploit weaknesses and vulnerabilities of networks, machines (fog nodes), and applications to get unauthorised access. It is challenging to provide security and privacy in the context of fog computing because existing mechanisms for cloud computing cannot be directly applied. Khan and Salah [8] surveyed security issues and their implications at each level (i.e., low level, intermediate, level, and high level) in the network. In the event of attacks, services are disturbed and the communication is interrupted. For instance, distributed denial-of-service (DDoS) constitutes one of the most serious security issues which concern all levels in the network. Vlajic and Zhou [9] report that IoT devices represent a fertile land for DDoS attacks; they propose blockchain as an enabling solution to counter these attacks. Recently, many studies [10–13] conclude that blockchain may be one of the suitable solutions to increase security when outsourcing and transmitting data from IoT devices to service providers.

Indeed, blockchain may be seen as a stunning innovation technology that revolutionises how we do business. This technology can be defined as a distributed digital ledger that keeps track of all the transactions that have taken place in a secure, chronological, and immutable way using peer-to-peer networking technology without the need for a trusted central entity. More specifically, blockchain technology is based on four main concepts: distributed ledgers, consensus mechanisms, cryptographic algorithms, and smart contracts. One of the first contributions aiming at securing a chain of blocks using cryptography is described in [14,15]; the authors designed a system where a document's timestamp cannot be changed. More importantly, blockchain ensures its users the possibility of transacting safely by reducing the security risks compared to current systems. It has many use in financial and non-financial applications as evidenced by the dramatic increase of interest by industry (e.g., IBM [16] and Microsoft [17]) and government. This technology has triggered a new wave of the economy and will considerably impact almost all industry segments which push it into the spotlight.

The concepts of nodes, computation, decentralisation, and distribution nature are held in common by blockchain and fog computing. In addition, both technologies are designated, in different research, to fulfil IoT requirements where blockchain is a key enabler to cope with security and privacy issues and fog computing is a key enabler to cope with computation, storage, and networking services. The main contribution of this chapter is to discuss the integration of blockchain with fog computing with the objective of securely sharing data and by preventing unauthorised users to access data. The remainder of this chapter is structured as illustrated in Figure 6.1.

6.2 CLOUD-TO-THING CONTINUUM BY MEANS OF FOG AND EDGE COMPUTING

Currently, the new distributed architecture promotes the emergence of smart devices and enables their networking and their deployment using 5G communication technologies. The use aggregates an exponential growth of data within the advancement of different types of applications ranging from smart transportation, e-health applications, retail, and other applications developed for robotics and augmented reality. Those devices are geospatially distributed; they generate and consume continuously huge amounts of data. With the aim of timely processing of this data, a new architecture was needed to extend cloud architecture. For this specific purpose, edge computing (enhanced data GSM environment) and fog computing (or mobile or virtual cloudlet cloud) are proposed to work together in a complementary way to improve the quality of services (QoS) and quality of experience (QoE) offered to consumers [18] (Figure 6.2).

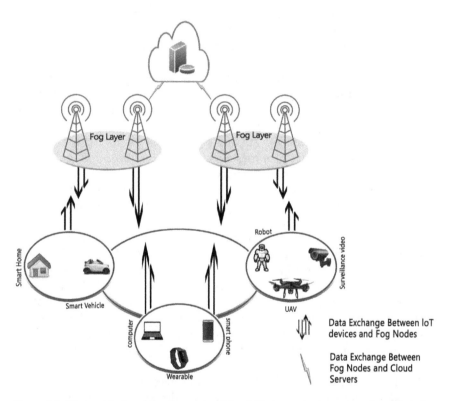

Figure 6.2 Hierarchical architecture (cloud-fog-IoT devices), the smart devices (vehicles, phones, robots, surveillance video, etc.) could offload their workloads on fog nodes located in near proximity on the edge which provides an intermediate layer between the devices layer and cloud layer. IoT, Internet of Things.

Fog nodes and edge nodes are deployed to ensure the interchange between different sources of data and to handle the massive handover of transactions in the network. Contrary to the centric architecture, the decentralised architecture is best suited to IoT applications and brings additional advantageous features, especially to mobile applications. After the data is offloaded to fog and edge nodes, computing, controlling, storage, and networking will be performed in real time. These nodes are not as robust as cloud nodes, but they are better than IoT devices. A critical data processing and a heavy or complicated computation task which requires a highly performing resources will be transferred to cloud computing. Also, the processing results of fog computing will be transferred into the cloud data centres for better data analytics.

Commonly, in the literature, edge computing and fog computing refer to the same paradigm. They are inclusive of a distributed architecture, close to the end devices, ensuring short delays. The OpenFog consortium puts the emphasis on the difference between these two concepts. Authors in [19] and [20] shed light on fog computing and edge computing by zooming in on their common and particular features of each one.

Starting with fog computing, Cisco was the pioneer to integrate this new paradigm [7]. It serves to ensure the distributions of its services horizontally in the network by optimizing the traversed route of data to be processed at the edge of the network. Moreover, this paradigm is considered a thread that leads to mutually connect data streams between the cloud and IoT devices. The OpenFog consortium proposes a flat architecture for pushing the computation resources close at the vicinity to the edge network. Furthermore, the agility of fog computing opens up a way to the developers to roll out services via application programming interfaces (APIs) and software development kits (SDKs).

On the other side, the edge computing paradigm was introduced by the European Telecommunication Standards Institute (ETSI) [21]. Basically, both paradigms are similar in several points in terms of architecture, flat nature, and functionalities with some differences. With these new concepts, processing data near to the location of devices shortens the response time compared to processing data in the cloud. Edge computing was a compelling achievement to accelerate the processing in real time for serving many applications and achieving a latency smaller than 1 ms [22].

Furthermore, fog computing has been conceived to control the intercommunication between edge devices and to push the computation, storage, and networking from the cloud to things, while the control of the intercommunication among end devices and the computation, storage, and networking functions are bordered at the local and edge network.

Table 6.1 compares the features of fog and edge computing:

Table 6.1 The characteristics of fog and edge computing

Features	Fog computing	Edge computing
Devices examples	Small servers, Switches, Access points	Edge routers, Base stations, Home gateways
Latency	Low	Low
Server nodes location	Between edge and cloud	Close to smart devices
Server nodes count	Fewer than edge servers	Very large
End-device mobility	Support mobility	Support mobility
Scalability	High	High
Interaction in real time	Timely processing	Timely processing
Heterogeneity	Yes	Yes
Architecture	Decentralised	Decentralised

6.2.1 Requirements

To realize the benefits of fog and edge computing, a number of requirements needs to be satisfied. Despite the advantages offered, the fog and edge infrastructure confront some challenges which give birth to some requirements to be addressed and resolved. The security and privacy requirements have been surveyed in [21,23,24]. The authors in [21] have identified the eight treat vectors (TVs) related to the edge computing layer. The IoT devices are the subject of several attacks such as physical tampering attack, hardware Trojan attack, side-channel attack, and softwarized attack (inject malicious code), which may impact negatively and create serious problems in the fog and edge nodes.

Table 6.2 summarises the security and privacy requirements of fog and edge computing.

Table 6.2 The requirements of fog and edge computing

Requirements	Definitions
Security	The security health of fog and edge nodes depends on three parameters, confidentiality, integrity, and availability. Basically, the communication is established in direct between end devices and fog and edge nodes. Primordially, the data must be accessible only by the eligible entities to ensure the confidentiality of data. For this purpose, it is quite important to establish authentication and access control mechanisms to fulfil the aforementioned requirement. Additionally, it is important to point out that data is replicated and shared among fog and edge devices which may result in data inconsistency, and loss of information. To avoid such a scenario, it is important to ensure data integrity. Furthermore, the necessity to offer uninterrupted services even when a fog or edge node breaks down. The services should be delivered by other nodes.
Privacy	Different privacy-preserving mechanisms are required to ensure the privacy of transactions and their partakers.

As mentioned before, blockchain technology was elected to be introduced with fog and edge computing in the attempt to provide security and privacy.

6.3 BLOCKCHAIN

Blockchain technology has redefined the concept of a traditional database and changed the way of controlling transactions carried out between different parties in the network. It offers a distributed ledgers which keep track of all transactions committed and where the control is performed by all the nodes interested in participating in the blockchain network. In essence, the intrinsic properties of blockchain are security and privacy established by its structure – secure by design – of cryptographic techniques used. Blockchain is a sequence of blocks linearly ordered over the time axis using the mechanism of timestamping which constitutes tamper-proof data blocks. The blocks are cryptographically bound through the hashing mechanisms which mainly maintain data integrity. Blockchain makes use of decentralised consensus mechanisms to ensure that all information recorded in the ledger are valid and verifiable; this is realised by achieving an agreement among the untrusted network nodes (i.e., nodes running the blockchain software or 'miners'). More specifically, when a user send a transaction (e.g., to transfer cryptocurrency) to the blockchain network, it is broadcast to all the network nodes. Upon receipt of the transaction, a node validates the transaction and, if validated, transmits the transaction to its neighbours. Periodically, a node is selected randomly to create a block that includes a list of valid transactions; the selection is implanted by the consensus mechanism used by the blockchain. Then, the node appends the new block to its copy of the blockchain and transmits this block to its neighbours. Upon reception of the new block, a node validates the block and, if validated, appends the block to its copy of the blockchain. Hence, the nodes come to an agreement about the true and valid state of the blockchain network assuming that at least 50% are not malicious.

When the Bitcoin application was premiered, it introduced a new concept based on miner nodes which are the censorship entities in the blockchain network. A principal mission attributed to miners is checking the validity of transactions and making a decision using a consensus algorithm with the goal to stand against double-spending attacks and strengthen the controls in the network. In fact, the more miner nodes participating in the network, the more security is established and more of the collusion of malicious nodes is decreased.

Blockchain technology has literally attracted attention in different areas. Figure 6.3 offers a glimpse of the fields of application as the authors surveyed in [25].

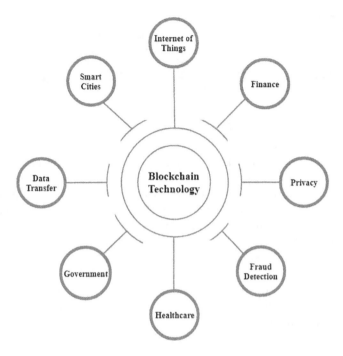

Figure 6.3 Some use scenarios of blockchain.

6.3.1 Transaction process

When a user wants to interact with the blockchain (e.g., to transfer cryptocurrency or store a testament), that user creates and signs, using a private key, a transaction; note that blockchain uses public-key encryption. Then, it sends the transaction to the blockchain network; a node that receives the transaction validates the transactions (e.g., verifies the user signature) and, if valid, stores the transaction in its pending list of transactions and transmits it to its neighbouring nodes. Periodically, a node is selected to create a block; the selection is based on the consensus protocol in use. The node then creates a block that consists of the block header and a list of transactions; the maximum number of transactions in a block depends on the blockchain that is used (e.g., bitcoin and Ethereum). The blockchain header includes (i) a timestamp that represents the time when the block is created; the timestamp is generated by secure mechanisms of timestamping to guarantee the chronological creation of blocks in the chain; (ii) a hash of the previous block; this makes blockchain immutable because it is not possible to modify any block without changing the entire chain. Indeed, changing even one bit in a block changes the value of its hash; (iii) a nonce which is an arbitrary number used for verify the validity of the block; this applies when proof-of-work (PoW) consensus mechanism is used; and (iv) root node of Merkle tree that represents the hash of all

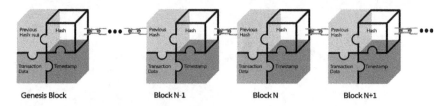

Figure 6.4 The data structure of blockchain.

transactions inside the block. The node that creates the new block appends it to its copy of the blockchain and sends it to its neighbouring nodes. Upon receipt of the new block, a node validates the block and, if valid, appends it to its copy of the blockchain and sends it to its neighbouring nodes. The block validation process includes (i) checking the Merkle tree's root hash; (ii) checking the validity of all transactions in the block; and (iii) checking the hash of the previous block. Broadcasting and validating the new block are what keep all the nodes on a network synchronised with each other.

A simplified structure of blockchain is illustrated in Figure 6.4: The blocks are cryptographically bound with a logical order. The fields of the block can vary according to the type of blockchain.

6.3.2 Types of nodes in the blockchain network

According to [26], there are three types of blockchain nodes: (i) mining nodes: they are responsible for producing blocks; each time a mining node produces a block, it is rewarded; (ii) full nodes: they are responsible for maintaining and distributing copies of the entire blockchain; in particular, they are responsible for the validation of blocks produced by the mining nodes. The more full nodes, the more decentralised the blockchain and the harder to hack; and (iii) light nodes: They perform similar functions as full nodes without maintaining the entire blockchain; in general, a light node connects to a full node, downloads only the headers of previous blocks, and connects to a full node. By way of illustrating, Figure 6.5 shows an example of blockchain operation; a transaction is generated by a light node and validated by miners. Then, a miner creates a new block, that includes the transaction, that is validated and appended to the blockchain by miners/full nodes.

6.3.3 Types of blockchain

Blockchain comes in many different types. More specifically, there are three types of blockchain [27]: permissionless, also known as a public blockchain (e.g., Bitcoin and Ethereum), permissioned blockchain, also known as consortium blockchain (e.g., hyperledger fabric), and a private blockchain. In public blockchain, any participant or user can write data to the blockchain and can read data recorded in the blockchain; anybody

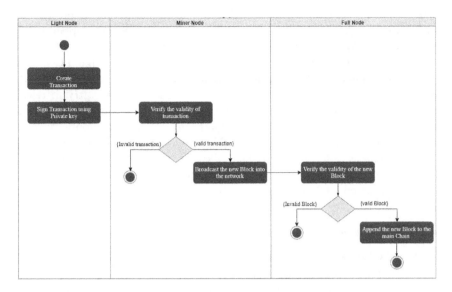

Figure 6.5 The transaction life cycle from its genesis to its integration to the main chain (Light node: creation step, mining node, and full node: verification step).

can be a full node, a miner, or a light node. Thus, there is little to no privacy for recorded data, and there are no regulations or rules for participants to join the network. Generally, a public blockchain is considered pseudo-anonymous (e.g., bitcoin and Ethereum); participants do not have to divulge their identity (e.g., name) instead they are linked to an address (i.e., the hash of the public key). Providing anonymity is difficult, but it is feasible (e.g., Zcash). The success of this type of blockchain depends on the number of participants, and it uses incentives to encourage more participation. Consortium blockchain put restrictions on who can participate. In particular, the creation and validation of blocks are controlled by a set of pre-authorised nodes; for example, we have a consortium of 10 banks where each bank operates one node. The right to read data recorded in the blockchain can be public or restricted to the participants. Even participants may be restricted on what they can do in the blockchain; for example, transactions between two participants may be hidden from the rest of the participants. In a private blockchain, write permissions are centralised and restricted to one entity; read permissions may be public or restricted.

6.3.4 Blockchain main ingredients

6.3.4.1 Consensus mechanisms

Consensus mechanisms are one of the key components of a blockchain. They are addressed for preventing and precluding the double spending and

Byzantine General (BG) problems [28]. All the distributed nodes present on the network have to follow and reach a consensus to successfully append new blocks to the blockchain. Consensus mechanisms can be categorised into two categories. Proof-based schemes such as POW and proof-of-stake (POS) [29] and the voting-based scheme such as Practical Byzantine Fault Tolerance [30]. In that way, the miners, by respecting the consensus protocol, try to reject the creation of illicit blocks undertaken by malicious entities. They operate differently, but they come together to reward miners as a recompose for their contribution which makes that competition takes place. The use of each consensus mechanism depends on the specificity of the blockchain applications, and they are yet under improvement with the aim to minimise the time and energy consumed while validating the created transactions [31].

6.3.4.2 Cryptography

One of the fundamental mechanisms of blockchain is cryptography. As we know, the etymology of cryptography has reference to the two Greek words 'cryptos graphein' that referred respectively to 'hidden write' [32]. In an insecure environment, the cryptography is introduced to entitle only the approved and accredited entities to extract data from the signed and encrypted messages using several techniques. Data integrity, data confidentiality, authentication, anonymity, and privacy preserving are the primary requirements achieved by cryptography. For the purpose of implementing a secure and private platform, the cryptographic algorithms are used in the blockchain platform to respond to the requirements aforementioned. Asymmetric cryptography or as also called 'public-key cryptography' has been mainly used to construct digital signature for signing transactions. Rivest-Shamir-Adleman (RSA) and Elliptic Curve Diffie-Hellman (ECDHA) are the most interesting cryptographic schemes used in the blockchain platform.

Cryptographic hashing [33] is a mechanism used to encrypt data transactions before broadcasting it in the network by its issuer. Secure hash algorithm (SHA256) is the famous hash algorithm implemented in several cryptosystems which relies on hashing an input parameter and converting it to an output with a size of 256 bit. It is a one-way function; that is, each input has one and only hash target as a result, and also there is no way to predict and to divine the input just based on the hash. Additionally, SHA-256d (SHA256 function is applied two times on data) is used for enhancing the complexity and the mining time (Bitcoin, Peercoin implement the SHA-256d). Also Scrypt (implemented by Litecoin) and EtHash (implemented by Ethereum) are other algorithms as a hashing function [34].

6.3.4.2.1 Signing transactions

Establishing the trust and reliability between different entities in the blockchain network is ensured by verifying whether the sender of the transaction is the real and the rightful owner of the data and whether the transaction data has been modified or altered in the network. The asymmetric cryptography bound to each entity in the network by a couple of keys (public key and private key). The public key is shared among all entities in the blockchain network, contrary to the private key which needs to be secretly kept from other entities regardless of the trust established in the network. The blockchain-based system has the autonomy to generate a unique address of 27-34 alphanumeric characters, referred to the global unique identifier (GUID), for each transaction using ECDSA or RSA with the secp256k1 [35]. According to [36], blockchain could allocate a number of addresses estimated to 1.46×10^{48} which enhances the addresses provided by IPv6. Each transaction has a sender and a receiver. The public key is used to generate an address associated with the transaction, using pay-to-pub key hash (P2PKH) method [35]. The private key is a 64-alphanumeric (256-bit hexadecimal format) used to sign the transactions, and it is undeniable which means that the issuers of the transaction cannot deny their ownership of data.

6.3.4.3 Smart contract

The concept of smart contract was introduced, before any blockchain technology was even in existence by nick Szabo in 1994. It defines the rules and penalties around an agreement in the same way that a traditional contract does, but also automatically enforce those obligations. A smart contract is a piece of code and data that is deployed to a blockchain; the first smart contract, with blockchain technology, was deployed on Ethereum in 2015. A smart contract can perform calculations, store information, and automatically send funds to other accounts. More specifically, it provides a number of functions/primitives (e.g., transfer token, store document, and buy token) that can be called by users. A smart contract is immutable (on blockchain); thus, it can be trusted by users to execute operations ranging from simple cryptocurrency transfers to more complex operations. On Ethereum, a smart contract can be implemented using Solidity language (Turing complete language) and deployed using RemixIDE [38].

6.3.5 Privacy

It is important to note that the transmission of data must be carried out in a private environment to ensure the non-disclosure of transaction metadata (sender and recipient addresses and the content). Hence, we can distinguish between two types of privacy, protecting the content of transactions and the identity of their partakers that are notably central requirements.

1. The relationships between the transactions and their issuers should be administered in the way as to not trace the real identity and to break off their linkability. The author in [36] indicates some analysis strategies

as anti-money launder (AML) or know your customer (KYC) that are applied in a public blockchain network and are able to pursue the pathway of transactions and extracting some essential information about the blockchain and responding to some important questions about who are using blockchain technology and for what purposes.

2. Privacy transaction refers to protecting the transaction value exchanged between different nodes in the network. Blockchain-based applications contain some critical information such as the medical history of patients, financial records, and so on which need to be safeguarded from malicious adversaries who have fraudulent or curiosity intents. This means that blockchain technology should manage the accessibility and the authentication of each entity.

The overall goal of blockchain is to meet the need for the requirements aforementioned by implementing different techniques to increase the level of privacy mentioned.

Among the different techniques used by blockchain is a Ring signature. The latter was firstly introduced by Rivest et al. [39] as a technique that can be used to cover up and conceal the real identity of transaction issuers and make them anonymous. It is based on the selection of a set of users to constitute a Ring from their public keys. The process is as follows: When users try to sign a transaction, they are addressed to use they private key and all the public keys of the ring (called 'RingSign' step). Then, the next step consists of verifying the validity of the transaction by a verifier (called 'RingVerify' step). With this technique, the issuer of a transaction is unreachable and untraceable. As explained in [36] and as depicted in Figure 6.6, users have a public key and a private key, respectively, and a ring is composed from a set of users' public keys. The users sign the message using all attributes in the ring. So, the verifier cannot predict whom the real owner of the message transaction for this situation because all the users are equiprobable.

Monero [40] is a distributed cryptocurrency application that implements CryptoNote which uses a ring signature and one-time-key to hide the origin and destination addresses of transactions. Later the scheme was upgraded and used the traceable ring signature for preventing the double-spending attacks and to make the content of the transactions anonymous.

Besides, a ring signature, another tool for enhancing privacy concerns in blockchain is Zero-Knowledge proof and is based on a prover P who needs to demonstrate to the verifier entity V that he possesses knowledge without revealing the original value. This mechanism is publicly verifiable, is used in several blockchain applications such as Zerocoin, and is usually used with smart contract execution. Furthermore, to controvert the analytics attack, the mixing techniques relying on a trustful intermediary for exchanging. This technique aims to break up the linkability between two addresses who transact usually with each other using a trusted address in-between instead

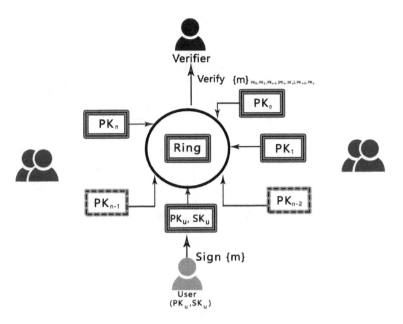

Figure 6.6 A ring signature inclusive of a set of public key of a group of users. The user performs the RingSign operation using all keys of the ring. Then, the verifier performs the RingVerify operations.

of transacting directly. Moreover, secure multi-party computation is a cryptographic scheme based on a group of participants who try to compute a function with keeping their private information secret. The homomorphic encryption is also used by blockchain to potentially enhance the privacy of its data by outsourcing the computation tasks without allowing the insourcing provider to predict anything about the value of the original data.

6.3.6 Blockchain characteristics

Table 6.3 summarises the potential features of blockchain focusing on immutability, traceability, synchronicity, anonymity, confidentiality, integrity, unicity, and simplicity.

6.3.7 Challenges

Despite all the benefits of blockchain has achieved in several areas, it also presents some problems. Decentralisation, security, and scalability are the major trilemmas of blockchain. Finding an agreement between these three interesting parameters is a real challenge [29]. As also argued by the authors in [41], blockchain present some drawbacks for its adoption with IoT.

Table 6.3 The characteristics of blockchain technology

Characteristics	Definition
Immutability	In a blockchain application, the records data are unforgeable and immutable. This is achieved by the mechanisms of timestamping by which the blocks are time-ordered in the chain and encryption and hashing which make the possibility of altering block data difficult, even more impossible.
Traceability	Proof-of-origin is a key to establish trust in the network. As a use of supply chain, the authors in [42] developed a platform called 'OriginChain' which offers traceability services of products using a consortium blockchain and a smart contract. Another use is social media. The authors in [43,44] are interested in using blockchain as a tool to trace the source of content news in social media to combat and struggle with fake news circulating on several channels of information.
Synchronicity	When a new block is successfully created and appended into the main chain, the updated version of blockchain will be simultaneously uploaded to all nodes in the network with respect to time.
Anonymity	Through the privacy mechanisms used by blockchain, the identity of a transaction, including its content and its partakers, is anonymous [45]
Confidentiality	Using public key cryptography in blockchain technology, the eligible instances are defined, as are the conditions under which users can access to participate in the blockchain network [46].
Integrity	Blockchain has been appraised for its integrated mechanisms of timestamping and the cryptographic signature, so the users can ensure that the life cycle of a transaction through a communication channel is safeguarded and managed by all the distributed ledgers.
Unicity	Using the consensus mechanism the double-spend attack is eliminated: A transaction cannot be performed twice or more.
Simplicity	The use of blockchain technology will simplify a set of operations. Also, this technology provides single registration [47], which means once the terminal device is registered, it can transact later without the need for new registration
Eliminate single point of failure	The blockchain nodes are distributed in a peer-to-peer network. And using a smart contract as an important ingredient of this technology, the blockchain is presented as a distributed manager [48].
Open source	Many blockchain applications such as Bitcoin, Ethereum, and Hyper Ledger are open-source platforms that help developers to base on to model their platform by comparing the properties of existing ones [45].

(Continued)

Table 6.3 (Continued) The characteristics of blockchain technology

Characteristics	Definition
Autonomy	Via rules predefined in smart contracts, the IoT devices are controlled and can behave autonomously [49].
Availability	One of the most desirous criteria of any application is the availability of data anywhere and at any time. The ledger of Blockchain is distributed and housed in every single node which affords it a high availability of data records on demand and permits it to avoid the single point of failure systems. Despite the malfunctioning of a node, the blockchain system operates correctly through the availability of the other nodes, and data is continuously available.
Privacy	The different techniques used by blockchain technology such as zero-knowledge-proof, ring signature, encryption, anonymisation, just to name a few, allow the communication without revealing the personal information of different entities in the network. This advantage matches with the GDPR keys. Many papers have already focused on access control using Blockchain technology for IoT applications. For example, the scheme proposed in [50] using FairAccess which integrates the PPDAC, operates equally as the RBAC and constitutes a decentralised framework which records the access control policies as transactions in the blockchain ledgers and achieves the anonymity and unlikability of users.
Accountability	In the case of selfish mining which engenders a fork problem, selfish miners are identified.
Reward	Grabbed the attention of nodes to participate by means of incentive mechanisms, a reward is earned by miners for their participation which creates an incentive mechanism.
Trustworthy	Using the consensus mechanism, the blockchain serves to establish trust in a distrusted environment.

Abbreviations: GDPR, General Data Protection Regulation; IoT, Internet of Things; PPDAC, privacy-preserving distributed access control; RBAC, role-based access control.

6.3.7.1 Scalability

The blockchain increasingly grows with the frequency of transactions launched continuously in the network. As blockchain widens in size, the devices with limited resources are inept and unqualified to deposit the whole copy of the chain on their system. Moreover, the mining process requires considerable time for execution. Hence, the throughput scales down. Actually, Bitcoin processes between 3 and 7 transactions per second, and Ethereum processes between 7 and 15 transactions, which are not suitable for the entire overwhelming transactions in the network and for the real-time applications. The author in [29] explains the unbalance between raising the block size and lightening the computation difficulty that will certainly increase the throughput but will diminish the level of security and open doors for attackers. Practically, blockchain requires an interval of time estimated to one hour to process data from its genesis to its integration into the main chain. As stated, Blockchain can only process a small number of transactions per second in front of the large number of transactions launched in the network, this constitutes a real challenge that should be surmounted.

6.3.7.2 Security

1. There are no perfect systems or algorithms that counterattack all types of vulnerabilities and avoid their serious effects. In this part, we highlight the several attacks threaten the Blockchain network and its participants. The authors in [45,51,52] zoom in on different attacks.
2. Blockchain can be subject to a Sybil attack, in which malicious participants are able to create forged peers in the network which behave badly. In a public blockchain, the mining process requires a high computation resource that represents a helpful method in which those attacks could be alleviated.
3. 51% of Sybil-like attacks, the malicious nodes have the influence and control over 51% of the mining pool in the network. It could endanger the trust of the data launched in transactions. For example, in a public blockchain, the dominated party may approve the creation of falsified and wrong transactions.
4. Certainly, the severe attack that frustrates the security of the blockchain is wallet theft. The blockchain participants safeguard their digital signature in a software known as a 'wallet' or a 'multi-signature wallet'. The hackers try to steal the information associated with each participant in the network. In a successful attempt, the hacker can impersonate the rightful owner and transact freely at a later time. Such an attack occurred in 2016 on the Bitfiex platform for exchanging Bitcoin in Honk Kong which underwent a wallet theft attack that resulted to a loss $72 million.
5. When Blockchain sets new policy and improves the version of its consensus protocol, a fork problem leads to distinguishing two types of nodes (the newer and older ones). Consequently and because if the versioning and incompatibility of nodes, the 'hard fork' and 'soft fork' can be generated and take place in such a situation. In both cases, the newer nodes having a great power in mining because they will respect the upgraded consensus rather than the old nodes.
6. Unfortunately, the traditional asymmetric cryptographic algorithms used by Blockchain become vulnerable in front of the malicious nodes armed with powerful quantum computers. However, in [53], the author proposes the post-quantum blockchain (PQB) as a solution to withstand against quantum computing attack.

6.3.7.3 Law

General Data Protection Regulation (GDPR) is European regulation. Its principal mission is to define and set up laws and restrictions to be applied by the European organisations in to protect sensitive personal information. We note in [54] that blockchain technology has some characteristics that seem contradictory to the GDPR principles. The immutability of data in blockchain and 'right to be forgotten' in the GPDR key. While the block is maintained into the blockchain, data could not be altered or deleted. By contrast, the 'right to be forgotten' consists that each entity should have the right to delete its own data (e.g., the health history of a patient) when it is necessary. The indelible nature of the data in the blockchain may present a block stumbling with law enforcement. And for this perspective, the authors in [55] have proposed a new approach acronym by FLPE (functionality-preserving local erasure) based on erasing data on a local level of nodes. Briefly put, this solution permits the clearing out the transactions in question while opening the doors for mischievous entities to transact freely.

6.4 BLOCKCHAIN AND FOG AND EDGE COMPUTING: A POSSIBLE COOPERATION

Definitely, fog computing has been considered the best solution for lightening and mitigating the pressure of requests launched by a large number of devices and by meeting the needs in low latency, real-time responsiveness, the high capability to handover the massive traffic, and better performances in computation tasks (Figure 6.7). With this new layer in the architecture, the computing services are decentred and the services are edged closer to end devices. However, users are worried and are reluctant about the services provided to the detriment of privacy and security of their sensitive data. At the same time, blockchain technology has been successfully reaching a robust platform with its security and privacy mechanisms [56]. For this purpose and to reinforce the communication in terms of security and privacy between the fog systems and terminal devices, the use of blockchain into fog nodes seems interesting. As described in [12], both technologies are made for each other and can promise a win-win situation. And for making a decision whether this technology is a valid solution to be implemented, a decision framework was introduced in [49]. Through the latter reference, the authors study the adequacy of blockchain technology in the environment of IoT and edge computing and define the parameters which help choose the appropriate platform (public, permissioned, or consortium). Many projects have conducted their research on the deployment of blockchain technology under fog computing which enables users to demand and receive fog services securely and privately. For instance, the

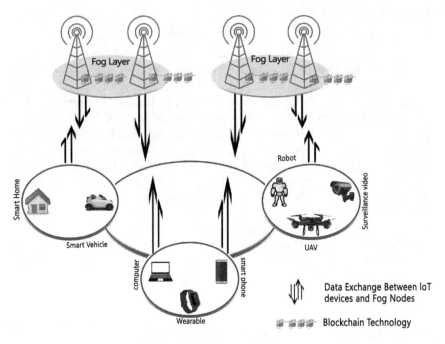

Figure 6.7 The Internet of Things (IoT) devices shared data with fog nodes. The latter nodes behave as a distributed ledgers for storage and for computation.

Industrial Edge Project [57] was launched in this perspective for industrial process and logistics management use. The authors in [58] point out that blockchain will serve to give surplus value to fog and edge infrastructure in terms of security and privacy which results in many studies to propose a template architecture combining both technologies.

6.4.1 Reference architecture

Many standards development organisations such as OpenFog Consortium and Industrial Internet Consortium have a major contribution to elaborate standards architecture, known as reference architecture, in edge computing as explained in [59]. The reference architecture is a template model on which the expert could to shape their own architecture. In the context of the H2020 project for factory automation, the authors have proposed a novel architecture FAR-EDGE reference architecture which has initiated the introduction of blockchain technology as the main component to protect data emanated from sensors [60]. Furthermore, the authors of [59] share the same vision as the latter reference by including blockchain technology to empower the security aspects in edge computing. They have conceptualised Global Edge Computing architecture as a reference architecture. Their approach outperforms the four existing reference architecture (i.e., the FAR-edge, the SAP Cloud Trust Center, the Edge Computing Consortium, and Industrial Internet Consortium) from which they have gotten inspired and taken advantage. Their proposed architecture verifies a set of important metrics including security and scalability. By the same token, [61] presents a conceptual reference architecture that supports a trusted orchestration management (TOM) for edge computing. The key role of the application of blockchain in this architecture is to ensure security properties, including the identification of hardware and software devices, mapping the data origin, and the non-repudiation with the aim to securely manage the orchestration information. In addition, the authors in [62] introduce HyperLedger as a control system for edge computing. This model is based on the IEC 61499 standard, which separates the architecture into two main levels, the supervisor level and the executive level. The first level is based on a Hyperledger fabric blockchain that distributes control and supervision of the executive level using smart contract with the aim of ensuring a secure transaction verification. The second level, the executive level, refers to edge nodes that deploy the docker containers to manage transactions issued from different devices. So all this architecture integrates blockchain on their architecture which enables secure transmission of data in the network.

6.4.2 Consensus mechanisms

Outsourcing the computational tasks into fog nodes is expected to satisfy a certain level of reliability and credibility between IoT devices and fog nodes. Bidirectional, fog nodes and IoT devices should be able to ensure the level of trustworthiness to collaborate and work together. On one hand, the terminals devices should ensure that the asked fog node could accomplish the computational tasks. On the other hand, the fog nodes need to ensure that the requestor device will pay the fees per computation service. Hence, establishing trust is one of the most significant aspects of the consensus mechanisms. As mentioned before, the traditional consensus mechanism such as PoW, PoS, and Practical Byzantine Fault Tolerance consumes a lot of time and energy while validating transactions. To answer to the scalability challenge of blockchain, new solutions are under development to reduce the appending waiting time and energy consumed for creating a new block by proposing new consensus algorithms and techniques for verification [58]. The authors in [63] propose a PoW consensus approach for fog computing using maximisation-factorisation statistics. The employed solution allows for solving the mathematical puzzle in PoW with less time and memory consumption. Also in the same regard and differently from the traditional consensus mechanisms such as (PoW, PoS, BFT, DPoS), the paper [64] introduces the proof-of-trust (PoT). The proposed consensus mechanism selects the entities who possess the higher level of trustworthiness as a miner known as 'Trust Blogger' in the TrustChain network. Among the features advantages of PoT consensus that there is no central authority that can control the network and the majority attack cannot take place in the TrustChain network. Another type of consensus mechanism applied in the context of edge computing is the proof-of-popularity (PoP) [30]. This mechanism relies on the knowledge wealth of edge nodes rather than their computing power. The evaluation of the adopted mechanism proof is that it enhances the level of security in edge environment.

6.4.3 Security advantages

It is commonly argued that data circulating on the IoT-fog-cloud infrastructure is the target of all types of attacks that plague all the layers in the network. Hence, the security and privacy of IoT data remain an important topic. The necessity for establishing trust and ensuring confidentiality, integrity, and availability are equally essential metrics that need to be addressed, especially for the fog and edge layer which constitute the bridge layer between the IoT and cloud layers. As stated in [46], the security services guided by Blockchain are inclusive of the aforementioned metrics. Seen in this perspective, a series of studies [21] have proposed blockchain as an approach to secure multi-access edge computing (MEC) by design. In a nutshell, trust is certainly a parameter that helps to retain users' confidence. With the same vision, BlockMEC [48] a blockchain-based MEC platform was proposed to establish a trustworthy environment between core nodes (edges nodes) and end devices using the services deployed by smart contracts. In the same perspective, the literature [65] introduces a payment platform based on Bitcoin, which aims to establish a trust between outsourcers and providers of the computational services (fog servers). Principally, the payment platform relies on the deposit mechanism and on smart contracts. The idea of this platform is to control the payment operations between the outsourcers and providers in a fair way. For example, when an outsourcer may not pay the computation fees for a reason, the provider will be automatically getting paid by deducting the amount from the deposit transaction of the outsourcer stored in the blockchain. Furthermore, the authors in [66] have proposed a mechanism by introducing blockchain technology to provide security services for lightweight devices by protecting them from untrusted edge servers. The authors in their proposed system introduce smart contracts to check the validity of edge servers. They opt for storing the validity states of edge services in the cloud nodes which helps IoT devices make decisions in choosing the appropriate edge device provider based on their reputation rating. In vehicular communication, the security of transmitted data constitutes the primordial parameter with the aim in maintaining the consistency of data and tracing its origin. The authors in [67] propose an architecture based on blockchain and edge computing. The roadside unit constitutes stable ledgers and are in charge of performing the mining operations and providing computation services. The critical information is stored in blockchain nodes to improve the level of security of the architecture. Furthermore and for standing against the scalability and security challenges of IoT devices, the authors in [68] designed a novel framework labelled EdgeChain based on edge computing, blockchain, and smart contracts. The proposed framework makes use of permissions blockchain technology and smart contracts to deliver security benefits such as immutability of data and to automate the resources allocation process by means of policies predefined on smart contracts. Blockchain technology is extended to the case of smart industrial processes. The authors of [69] try to investigate the different attacks and treat those that would jeopardise each layer in the network. In this sense, the authors propose SEC-BlockEdge for Log-HOUSE Construction using blockchain and edge computing as an effective solution to enhance the performance of industrial process. Their studies indicate that the ledger layer contributes to add valuable advantages to each layer in the network. And especially for the edge layer, it contributes and helps in processing, storing, and sharing data coming from different places securely and privately. Principally, blockchain technology provides a tamper-proof mechanism for transaction records. It was the solution proposed by different researchers to enhance the integrity of the data on the fog nodes by preventing any amendments by malicious parties. Reference [70] comes with a blockchain-enabled computation offloading method labelled BeCome aimed to achieve the integrity of data in MEC in the meantime optimizing the edge computing devices tasks and the offloading time and energy. In the same case of cloud computing, [71] has proposed blockchain as a database system based on the distributed replicated ledger which is able to ensure integrity in the cloud computing paradigm. Furthermore, to stand against fraudulent activities in the network, the authors in [72] designed a novel framework called Cochain-SC that articulates principally on the Ethereum blockchain using smart contract and SDN. Through the framework aforementioned, the authors aim to mitigate and counter the DDoS attack and to detect the bad demeanour of attackers in real time.

6.4.4 Privacy advantages

For better data privacy, it would be advantageous to transfer data to the decentralised systems rather than the centralised systems for not caring about the single point of failure and to minimise the dangers. For this purpose, IoT devices delegate and offload their data to be computed in the fog nodes. Hence, maintaining the privacy of the value traded is paramount of importance. Blockchain technology sets up a panoply of techniques that give issuers of transactions the possibility to control their anonymity by hiding the address of the sender, receiver, and the content of the transactions [73]. So, deploying the blockchain on fog nodes will increase the level of privacy. It is equally important to mention that blockchain permits the approved entities to belong to the peer's network by providing them an identity certificate, and it has also the control to exclude the untrusted entities from its network. To this end, the BSeIn framework was proposed in [47] and makes use of cryptographic schemes (Attribute-Based Signatures, Multi-Receivers Encryption, Advanced Encryption Standard, Message Authentication Code) to ensure anonymous authentication, auditability, confidentiality, and fine-grained access control in fog-computing layer. Furthermore, [64] introduces a novel privacy-preserving blockchain with edge computing called TrustChain. Principally, TrustChain is designed to keep personal information secret by implementing zero-knowledge-proof, encryption, and anonymisation as techniques that permit users to interact with service providers privately. An additional feature of the latter framework is permitting users to delete data by using a cryptographic link which matches a GDPR key (the right to the erasure). For the same case, a user-centric blockchain (UCB) is proposed for preserving copyright and edge knowledge [30] using blockchain. Another applicable use is the carpooling systems that presents an effective solution to coordinate between a group of people (passengers and drivers) who share the same destination and the same itinerary route to partake the same vehicle. The authors in [74] have proposed an efficient and privacy-preserving carpooling model called 'FICA' using blockchain-assisted vehicular fog computing to avoid privacy leakage of data relative to passengers and drivers that expose their sensitive information such as their real location and their destination. The FICA makes use of a private blockchain and implements an anonymous authentication scheme, private proximity test, and privacy-preserving range query scheme to meet and assure a set of objectives and a set of security parameter including (i) authentication, the scheme allows an anonymous authentication to users and permit to fog nodes to verify their real location; (ii) data confidentiality and integrity, the carpooling records are protected in the private blockchain network from any external or internal alteration; (iii) user privacy, the identity of users is preserved; (iv) traceability, the trusted authority could track the identity of a user in the case of a bad demeanour; and (v) efficiency, the model introduces blockchain to lessen the cost of storage and introduces fog nodes to lessen the cost of computational tasks. The literature [75] brings another use for energy trading to establish security and privacy while the distribution of electric energy between different electric vehicles. A secure mechanism for payment was established. The idea was to integrate a framework based on consortium Blockchain and edge computing. The principal actors in the framework are: (i) electric vehicles (EV), the IoT nodes; they can share their energy in a peer-to-peer network to maintain the equilibrium of energy between supply and demand, (ii) authorized local energy aggregators (LEAGs) are the authority that distributes the digital signature to EV in the network and audits the transaction validity, and (iii) edge service provider (ESP) represent the edge nodes. The identity is preserved through the public key associated with each EV to prevent an attacker from mapping the trajectory and situating a specific EV through to a real identity. The computation services are purchased from the ESP by the LEAGs to successfully create the transactions in the blockchain and enhance the scalability in the network.

6.4.5 Resource management using blockchain

Among the key roles of fog computing is distributing the computational tasks. As mentioned in the deployment of blockchain-assisted IoT devices are challenged with the three reasons: (i) the quite limited capabilities resources which are usually not qualified to execute the computation tasks locally, (ii) the limited bandwidth, and (iii) the need also to maintain their power. The authors in [76] experimented and demonstrated that fog computing is potentially suited to host Blockchain applications rather than cloud computing. The solution is to delegate the process of mining to a close node (fog node) in the network. Hence, this delegation outperforms the computation tasks but lacks dispatching the priority of the IoT nodes' resources. Accordingly, with the high number of requests per device, the traffic will be congested. A proper framework based on permissioned Blockchain was introduced as credit-based resource management to administer the resources allocation of IoT devices continuum and automating the administration is the key role of the smart contract [77]. In [78], the author has proposed a pricing mechanism to mitigate congestion and supporting mobile devices in mining tasks. As such, the price of services will fluctuate according to the supply and demand. The Stackelberg game theory was applied to manage the resources of edge systems and mobile devices. Blockchain and fog computing need to cope with the heterogeneity of different end users and take into consideration the possibility of giving them the leeway to choose the appropriate nodes units for their needs [79].

6.4.6 Challenges of the incorporation

Undeniably, blockchain technology has fruitful advantages in terms of security and privacy services. However, the integration of both technologies, fog and edge computing with blockchain, also presents some issues and restrictions. The studies in this area are still in its infancy, which underscored the importance of accelerating researches in this direction to achieve certain level of maturity. As a first limitation, the traditional industries cannot easily opt for new technology such as blockchain. They need to analyse in depth whether the integration of both will be a good solution that fits their objectives [42]. The second limitation is the high cost of money and time will curb the integration of blockchain in several uses. For instance, the authors share with us their learned lessons about their experiment with OriginChain. An additional limitation is mobility. Practically, most of the researches do not take into consideration the mobility aspect of user devices, for instance, the high-speed of vehicles [67], which may engender a serious problem of instability [80], and error-prone wireless transmission links [81], which is not adequate with blockchain ledgers.

6.5 FUTURE DIRECTIONS

We hope through this chapter that we have provided a clear view about the incorporation of blockchain with fog and edge computing from the security and privacy standpoints. In the future lines of research, we will concentrate and deepen our study to survey the consensus and incentivisation mechanisms, reference architecture, and privacy techniques appropriate to fog and edge computing.

6.6 CONCLUSIONS

In the ensuing years, blockchain technology will capture a considerable interest and bring a valuable contribution to different areas. More precisely with its security and privacy mechanisms, it is voted to be applied in the case of the fog and edge computing concepts. In this chapter, we have outlined the surplus value of this technology. It is considered a key enabler of security and privacy by verifying publicly the validity of transactions, establishing trust, preserving the privacy of the transactions and their partakers, self-executing transactions, collecting validated information, and preventing and standing against destructive attacks. With all of the latter advantageous features, blockchain is ideally the suited solution to open up many opportunities to fog and edge computing layer and to reinforce and strengthen the immunity of fog and edge systems.

REFERENCES

1. E. Anzelmo, A. Bassi, D. Caprio, S. Dodson, R. Van Kranenburg, and M. Ratto, Discussion Paper on the Internet of Things commissioned by the Institute for Internet and Society, Berlin, pp. 1–61, 2011.
2. G. Knieps, 'Internet of Things, big data and the economics of networked vehicles,' *Telecomm. Policy*, vol. 43, no. 2, pp. 171–181, 2019.
3. J. S. Walia, H. Hämmäinen, K. Kilkki, and S. Yrjölä, '5G network slicing strategies for a smart factory,' *Comput. Ind.*, vol. 111, pp. 108–120, 2019.
4. F. Bonomi, Connected vehicles, the Internet of Things, and fog computing. In *The Eighth ACM International Workshop on Vehicular Inter-Networking (VANET)*, Las Vegas, NV.
5. 'OpenFog Consortium.' [Online]. Available: www.openfogconsortium.org.
6. T. H. Ashrafi, M. A. Hossain, S. E. Arefin, K. D. J. Das, and A. Chakrabarty, 'Service based FOG computing model for IoT,' *Proc. 2017 IEEE 3rd Int. Conf. Collab. Internet Comput. CIC 2017*, vol. 2017-January, pp. 163–172, 2017.

7. Cisco, 'Fog Computing and the Internet of Things: Extend the Cloud to Where the Things Are.'

8. M. A. Khan and K. Salah, 'IoT security: Review, blockchain solutions, and open challenges,' *Futur. Gener. Comput. Syst.*, vol. 82, pp. 395–411, 2018.

9. N. Vlajic and D. Zhou, 'IoT as a land of opportunity for DDoS hackers,' *Computer (Long. Beach. Calif).*, vol. 51, no. 7, pp. 26–34, 2018.

10. R. Kuhn and T. Weil, 'Can blockchain strengthen the IoT?,' *Secur. IT*, pp. 68–72, 2017.

11. V. Dedeoglu, R. Jurdak, G. D. Putra, A. Dorri, and S. S. Kanhere, 'A Trust Architecture for Blockchain in IoT,' 2019.

12. 'No Title.' [Online]. Available: https://blogs.cisco.com/innovation/blockchain-and-fog-made-for-each-other.

13. Q. Shafi and A. Basit, 'DDoS botnet prevention using blockchain in software defined Internet of Things,' *Proc. 2019 16th Int. Bhurban Conf. Appl. Sci. Technol. IBCAST 2019*, pp. 624–628, 2019.

14. S. Haber and S. Stornetta, 'Introduction,' *J. Cryptol.*, vol. 3, no. 2, pp. 99–111, 1991.

15. W. Bolt, 'Bitcoin and cryptocurrency technologies: A comprehensive introduction,' *J. Econ. Lit.*, vol. 55, no. 2, pp. 467–469, 2017.

16. 'IBM.' [Online]. Available: www.ibm.com/fr-fr/blockchain.

17. 'Microsoft.' [Online]. Available: https://azure.microsoft.com/en-us/solutions/blockchain/.

18. T. Taleb, S. Dutta, A. Ksentini, M. Iqbal, and H. Flinck, 'Mobile edge computing potential in making cities smarter,' *IEEE Commun. Mag.*, vol. 55, no. 3, pp. 38–43, 2017.

19. A. Yousefpour et al., 'All one needs to know about fog computing and related edge computing paradigms: A complete survey,' *J. Syst. Archit.*, 2019.

20. T. Internet, 'Tutorial 1: What is fog computing and how is it different from edge computing?,' no. April, pp. 18–20, 2017.

21. P. Ranaweera, A. D. Jurcut, and M. Liyanage, '9. Realizing multi-access edge computing feasibility: Security perspective,' no. September, 2019.

22. M. Chiang and T. Zhang, 'Fog and IoT: An overview of research opportunities,' *IEEE Internet Things J.*, vol. 3, no. 6, pp. 854–864, 2016.

23. S. Raza, S. Wang, M. Ahmed, and M. R. Anwar, 'A survey on vehicular edge computing: Architecture, applications, technical issues, and future directions,' *Wirel. Commun. Mob. Comput.*, vol. 2019, 2019.

24. P. K. Sharma, M. Y. Chen, and J. H. Park, 'A software defined fog node based distributed blockchain cloud architecture for IoT,' *IEEE Access*, vol. 6, no. c, pp. 115–124, 2018.

25. J. Abou Jaoude and R. George Saade, 'Blockchain applications: Usage in different domains,' *IEEE Access*, vol. 7, pp. 45360–45381, 2019.

26. I. Makhdoom, M. Abolhasan, H. Abbas, and W. Ni, 'Blockchain's adoption in IoT: The challenges, and a way forward,' *J. Netw. Comput. Appl.*, vol. 125, no. October 2018, pp. 251–279, 2019.

27. D. Mingxiao, M. Xiaofeng, Z. Zhe, W. Xiangwei, and C. Qijun, 'A review on consensus algorithm of blockchain,' *2017 IEEE Int. Conf. Syst. Man, Cybern. SMC 2017*, vol. 2017-January, pp. 2567–2572, 2017.

28. K. Wust and A. Gervais, 'Do you need a blockchain?,' *Proc. 2018 Crypto Val. Conf. Blockchain Technol. CVCBT 2018*, no. i, pp. 45–54, 2018.

29. R. Yang, F. R. Yu, P. Si, Z. Yang, and Y. Zhang, 'Integrated blockchain and edge computing systems: A survey, some research issues and challenges,' *IEEE Commun. Surv. Tutorials*, vol. PP, no. c, pp. 1–1, 2019.

30. G. Li, M. Dong, L. T. Yang, K. Ota, J. Wu, and J. Li, 'Preserving edge knowledge sharing among IoT services: A blockchain-based approach,' no. July, 2019.

31. Z. Zheng, S. Xie, H. Dai, X. Chen, and H. Wang, 'An overview of blockchain technology: Architecture, consensus, and future trends,' *Proc. 2017 IEEE 6th Int. Congr. Big Data, BigData Congr. 2017*, pp. 557–564, 2017.

32. 'Cryptography.' [Online]. Available: www.etymonline.com.

33. National Institute of Standards and Technology (NIST), 'Secure Hash Standard (SHS) (FIPS PUB 180-4),' *Fed. Inf. Process. Stand. Publ.*, vol. 180–4, no. August, p. 36, 2015.

34. T. M. Fernández-Caramés and P. Fraga-Lamas, 'A review on the use of blockchain for the Internet of Things,' *IEEE Access*, vol. 6, no. c, pp. 32979–33001, 2018.

35. N. Z. Aitzhan and D. Svetinovic, 'Security and privacy in decentralized energy trading through multi-signatures, blockchain and anonymous messaging streams,' *IEEE Trans. Dependable Secur. Comput.*, vol. 15, no. 5, pp. 840–852, 2018.

36. Q. Feng, D. He, S. Zeadally, M. K. Khan, and N. Kumar, 'A survey on privacy protection in blockchain system,' *J. Netw. Comput. Appl.*, vol. 126, pp. 45–58, 2019.

37. A. H. L. Luu, D.-H. Chu, H. Olickel, P. Saxena, 'Making smart contracts smarter,' *Proc. 2016 ACM SIGSAC Conf. Comput. Commun. Secur.*, vol. 28, no. 6–7, pp. 254–269, 2016.

38. F. Schrans, S. Eisenbach, and S. Drossopoulou, 'Writing safe smart contracts in flint,' *ACM Int. Conf. Proceeding Ser.*, vol. Part F1376, no. June, pp. 218–219, 2018.

39. R. L. Rivest, A. Shamir, and Y. Tauman, 'How to leak a secret: Theory and applications of ring signatures,' pp. 164–186, 2006.

40. 'Monero.' [Online]. Available: ttps://web.getmonero.org/.

41. M. S. Ali, M. Vecchio, M. Pincheira, K. Dolui, F. Antonelli, and M. H. Rehmani, 'Applications of blockchains in the Internet of Things: A comprehensive survey,' *IEEE Commun. Surv. Tutorials*, vol. 21, no. 2, pp. 1676–1717, 2019.

42. Q. Lu and X. Xu, 'Adaptable blockchain-based systems: A case study for product traceability,' *IEEE Softw.*, vol. 34, no. 6, pp. 21–27, 2017.

43. A. Qayyum, J. Qadir, M. U. Janjua, and F. Sher, 'Using blockchain to rein in the new post-truth world and check the spread of fake news,' 2019.

44. P. Fraga-Lamas and T. M. Fernández-Caramés, 'Leveraging distributed ledger technologies and blockchain to combat fake news,' 2019.

45. I. C. Lin and T. C. Liao, 'A survey of blockchain security issues and challenges,' *Int. J. Netw. Secur.*, vol. 19, no. 5, pp. 653–659, 2017.

46. T. Salman, M. Zolanvari, A. Erbad, R. Jain, and M. Samaka, 'Security services using blockchains: A state of the art survey,' *IEEE Commun. Surv. Tutorials*, vol. 21, no. 1, pp. 858–880, 2019.

47. C. Lin, D. He, X. Huang, K. K. R. Choo, and A. V. Vasilakos, 'BSeIn: A blockchain-based secure mutual authentication with fine-grained access control system for industry 4.0,' *J. Netw. Comput. Appl.*, vol. 116, no. May, pp. 42–52, 2018.

48. Y. Zhu, 'A survey on mobile edge platform with blockchain,' *Proc. 2019 IEEE 3rd Inf. Technol. Networking, Electron. Autom. Control Conf. ITNEC 2019*, no. Itnec, pp. 879–883, 2019.
49. C. Pahl, N. El Ioini, and S. Helmer, 'A decision framework for blockchain platforms for IoT and edge computing,' *IoTBDS 2018: Proc. 3rd Int. Conf. Internet Things, Big Data Secur.*, vol. 2018-March, pp. 105–113, 2018.
50. A. Ouaddah, *A Blockchain Based Access Control Framework for the Security and Privacy of IoT with Strong Anonymity Unlinkability and Intractability Guarantees*, 1st ed. Elsevier, 2018.
51. J. C. Song, M. A. Demir, J. J. Prevost, and P. Rad, 'Blockchain design for trusted decentralized IoT networks,' *2018 13th Syst. Syst. Eng. Conf. SoSE 2018*, pp. 169–174, 2018.
52. W. Gao, W. G. Hatcher, and W. Yu, 'A survey of blockchain: Techniques, applications, and challenges,' *Proc. Int. Conf. Comput. Commun. Networks, ICCCN*, vol. 2018-July, no. i, 2018.
53. Y.-L. Chen, Y.-X. Yang, X.-B. Chen, X.-X. Niu, Y. Sun, and Y.-L. Gao, 'A secure cryptocurrency scheme based on post-quantum blockchain,' *IEEE Access*, vol. 6, no. Part Ii, pp. 27205–27213, 2018.
54. C. Alexopoulos, Y. Charalabidis, A. Androutsopoulou, M. A. Loutsaris, and Z. Lachana, 'Benefits and obstacles of blockchain applications in e-government,' *Proc. 52nd Hawaii Int. Conf. Syst. Sci.*, pp. 3377–3386, 2019.
55. M. Florian, S. Beaucamp, S. Henningsen, and B. Scheuermann, 'Erasing data from blockchain nodes,' 2019.
56. A. Prashanth Joshi, M. Han, and Y. Wang, 'A survey on security and privacy issues of blockchain technology,' *Math. Found. Comput.*, vol. 1, no. 2, pp. 121–147, 2018.
57. 'Industrial Edge Project.' [Online]. Available: www.oulu.fi/cwc/ industrialedge.
58. M. H. Ziegler, M. Grobmann, and U. R. Krieger, 'Integration of fog computing and blockchain technology using the plasma framework,' pp. 120–123, 2019.
59. I. Sittón-Candanedo, R. S. Alonso, J. M. Corchado, S. Rodríguez-González, and R. Casado-Vara, 'A review of edge computing reference architectures and a new global edge proposal,' *Futur. Gener. Comput. Syst.*, vol. 99, no. 2019, pp. 278–294, 2019.
60. J. Soldatos, 'Combining edge computing and blockchains for flexibility and performance in industrial automation,' no. c, pp. 159–164, 2017.
61. C. Pahl, N. El Ioini, S. Helmer, and B. Lee, 'An architecture pattern for trusted orchestration in IoT edge clouds,' *2018 3rd Int. Conf. Fog Mob. EdgeComput. FMEC 2018*, pp. 63–70, 2018.
62. A. Stanciu, 'Blockchain based distributed control system for edge computing,' *Proc. 2017 21st Int. Conf. Control Syst. Comput. CSCS 2017*, pp. 667–671, 2017.
63. G. Kumar, R. Saha, M. K. Rai, R. Thomas, and T.-H. Kim, 'Proof-of-work consensus approach in blockchain technology for cloud and fog computing using maximization-factorization statistics,' *IEEE Internet Things J.*, vol. 6, no. 4, pp. 6835–6842, 2019.
64. U. Jayasinghe, G. M. Lee, Á. MacDermott, and W. S. Rhee, 'TrustChain: A privacy preserving blockchain with edge computing,' *Wirel. Commun. Mob. Comput.*, vol. 2019, pp. 1–17, 2019.

65. H. Huang, X. Chen, Q. Wu, X. Huang, and J. Shen, 'Bitcoin-based fair payments for outsourcing computations of fog devices,' *Futur. Gener. Comput. Syst.*, vol. 78, pp. 850–858, 2018.

66. M. Rehman, N. Javaid, M. Awais, M. Imran, and N. Naseer, 'Cloud based secure service providing for IoTs using blockchain,' no. July, 2019.

67. X. Zhang, R. Li, and B. Cui, 'A security architecture of VANET based on blockchain and mobile edge computing,' *Proc. 2018 1st IEEE Int. Conf. Hot Information-Centric Networking, HotICN 2018*, no. 201702019, pp. 258–259, 2019.

68. J. Pan, J. Wang, A. Hester, I. Alqerm, Y. Liu, and Y. Zhao, 'EdgeChain: An edge-IoT framework and prototype based on blockchain and smart contracts,' *IEEE Internet Things J.*, vol. 6, no. 3, pp. 4719–4732, 2019.

69. T. Kumar, A. Braeken, V. Ramani, I. Ahmad, E. Harjula, and M. Ylianttila, 'SEC-BlockEdge: Security threats in blockchain-edge based industrial IoT networks,' no. July, 2019.

70. X. Xu, X. Zhang, H. Gao, Y. Xue, L. Qi, and W. Dou, 'BeCome: Blockchain-enabled computation offloading for IoT in mobile edge computing,' *IEEE Trans. Ind. Informatics*, vol. PP, no. X, pp. 1–1, 2019.

71. E. Gaetani, L. Aniello, R. Baldoni, F. Lombardi, A. Margheri, and V. Sassone, 'Blockchain-based database to ensure data integrity in cloud forensics,' *Int. J. Sci. Res. Comput. Sci. Eng. Inf. Technol.* © 2018 IJSRCSEIT, vol. 4, no. 10, pp. 2456–3307, 2018.

72. Z. Abou El Houda, A. S. Hafid, and L. Khoukhi, 'Cochain-SC: An intra- and inter-domain DDoS mitigation scheme based on blockchain using SDN and smart contract,' *IEEE Access*, vol. 7, pp. 98893–98907, 2019.

73. G. Zyskind, O. Nathan, and A. S. Pentland, 'Decentralizing privacy: Using blockchain to protect personal data,' *Proc. 2015 IEEE Secur. Priv. Work. SPW 2015*, pp. 180–184, 2015.

74. M. Li, L. Zhu, and X. Lin, 'Efficient and privacy-preserving carpooling using blockchain-assisted vehicular fog computing,' *IEEE Internet Things J.*, vol. PP, no. AUGUST, p. 1, 2018.

75. Z. Zhou, L. Tan, and G. Xu, 'Blockchain and edge computing based vehicle-to-grid energy trading in energy Internet,' *2nd IEEE Conf. Energy Internet Energy Syst. Integr. EI2 2018: Proc.*, pp. 1–5, 2018.

76. M. Samaniego and R. Deters, 'Blockchain as a service for IoT,' *Proc. 2016 IEEE Int. Conf. Internet Things; IEEE Green Comput. Commun. IEEE Cyber, Phys. Soc. Comput. IEEE Smart Data, iThings-GreenCom-CPSCom-Smart Data 2016*, pp. 433–436, 2017.

77. J. Pan, J. Wang, A. Hester, I. Alqerm, Y. Liu, and Y. Zhao, 'EdgeChain: An edge-IoT framework and prototype based on blockchain and smart contracts,' *IEEE Internet Things J.*, vol. PP, no. c, p. 1, 2018.

78. Z. Xiong, S. Feng, D. Niyato, P. Wang, and Z. Han, 'Optimal pricing-based edge computing resource management in mobile blockchain,' *IEEE Int. Conf. Commun.*, vol. 2018-May, 2018.

79. M. Conoscenti and J. Carlos De Martin, 'IOT_blockchain for the Internet of Things: A systematic literature review,' *Third Int. Symp. Internet Things Syst. Manag. Secur.*, pp. 1–6, 2016.

80. S. Kim, 'Impacts of mobility on performance of blockchain in VANET,' *IEEE Access*, vol. 7, no. 1, pp. 68646–68655, 2019.
81. T. Jiang, H. Fang, and H. Wang, 'Blockchain-based Internet of Vehicles: Distributed network architecture and performance analysis,' *IEEE Internet Things J.*, vol. 6, no. 3, pp. 4640–4649, 2019.

Chapter 7

Scaling edge computing security with blockchain technologies

Pau Marcer, Xavier Masip, Eva Marin, and Alejandro Jurnet

CONTENTS

7.1 EDGE SECURITY CHALLENGES

7.1.1 Introducing security at the edge

Cloud computing refers to centralised services running on homogeneous clusters, usually controlled and maintained by a single organisation (this may change when deploying multi-cloud services though), and consequently driving to a similar performance. On a different basis, edge computing conceptually refers to highly decentralised systems supporting services execution throughout heterogeneous clusters of devices deployed at the edge, each one probably embedding different set of capabilities, thus turning into distinct levels of performance.

The significant differences shown between cloud and edge paradigms are highly emphasised when addressing security. Focusing on the cloud, because services are centralised, the security may be also handled in a centralised way with no relevant impact on the whole system performance. That said, it is widely accepted that centralised systems suffer from the single-point-of-failure problem (i.e., should the centralised security server or network fails, so will the cloud service or network). However, centralised systems are managed with fault-proof networks replicated through the whole infrastructure, and what is referred to as the high availability property offered by cloud providers makes errors on that side seldom and most errors are motivated by the services running at cloud infrastructure.

Nevertheless, the whole scenario is different at the edge. Indeed, many notable assumptions and constraints are worthy to be considered, namely three areas: connectivity, scalability, and mobility. First, edge devices are not always connected, what draws a scenario rather different from cloud (e.g., drivers for this assertion may be a network fail, on/off energy savings policies, etc.). Therefore, should edge devices want to rely on the cloud to be secure, the non-guaranteed connectivity would also drive a non-guaranteed security. Consequently, the edge should never rely on the cloud to be secure, rather the cloud should be a complementary tool to secure the edge and never the core of that security.

The second refers to the scalability. It is widely known that the cloud can tolerate billions of connections because most cloud networks are designed to do so. However, recognition of the fact that when moving towards the edge, the network is dimensioned conceptually different, a large amount of connections at the edge might overload the edge network acting as a potential denial-of-service (DoS) to its devices. In addition because user profiles are usually stored in centralised servers at cloud, scaling up users (at the edge) will consequently trigger scalability issues at the edge network.

Third, mobility brings in a non-negligible set of issues to the edge that do not substantially affect the cloud. Indeed, a huge portion of the most used security models, such as certification authorities (CAs) and certificates are used to provide security to static services (i.e., webpages, webservers) and operate on static domains. Certainly, this strategy is useless when shifting to the edge, where device mobility is a must. Then, according to the current CAs and certificates models, a new certificate will have to be issued wherever a device changes its Internet protocol (IP) address (in mobile scenarios, according to currently deployed IP allocation strategies would be the usual procedure). This will drive severe scalability issues as CAs will be overloaded with requests from a whole set of devices on the move. This issue will be exacerbated when considering the connection facilities (i.e., large deployment of antennas) to be brought in by a wide adoption of the 5G technology.

Aimed at summarizing the envisioned edge scenario, Figure 7.1 illustrates the scenario considered in this chapter, which consists of an edge network, composed by clusters of Internet of Things (IoT) devices, connected to the cloud through Internet. In this scenario, we argue that security solutions based solely on cloud are not feasible.

In summary, the edge paradigm requires an innovative security architecture designed to be as flexible as the edge would be, while at the same time scaling up similar to cloud. The main aim of this chapter is to discuss on the challenges to be faced towards securing the set of edge elements and clusters of edge devices within the edge network. The main contributions in this chapter are:

- Analysing the main challenges when securing the edge scenario.
- Analysing the possible problems encountered when applying blockchain over the edge.

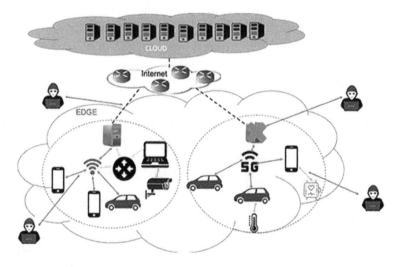

Figure 7.1 Envisioned edge scenario.

- Proposing a blockchain model to be used in the edge scenario for decentralised authentication and validation of the profile data provided by the edge devices.
- Presenting the main use of the proposed model.
- Presenting a series of cases comparing the proposed model to cloud solutions.

The chapter is organised as follows. The subsections in the Introduction will describe the architectural challenges when trying to secure the edge, as well as the unfeasibility of securing the edge using cloud computing. Section 7.2 sets the rationale for using blockchain in the security field. Section 7.3 is devoted to review the main concepts in blockchain. Section 7.4 presents a tentative blockchain architecture intended to provide decentralised authentication and data validation (edge devices data) to a set of edge devices. Section 7.5 compares the proposed blockchain solution with the existing cloud technologies.

7.1.2 Edge architectural challenges

In this section the main architectural challenges encountered at the edge are introduced and overviewed, most of them related to the security limitations imposed by a potential use of traditional cloud models at the edge. Three key challenges are analysed: (i) scalability, describing the main issues coming up when merging cloud and edge; (ii) the imposed security limitations to be faced at the edge when working with cloud solutions; and (iii) the connection dependency for some edge devices to be secure, motivated by the limitations inherent to constrained edge devices that make them to rely on other systems they should be connected to guarantee security provisioning.

7.1.2.1 Scalability

Scalability is indeed an important issue in edge computing. The rationale behind this assertion is mainly rooted on the vast amount of heterogeneous and on-the-move devices the edge brings together. This context makes cloud centralised strategies not good enough because both the network and the offered services may be easily collapsed by the expected number of requests generated by devices at the edge.

Moreover, it is worth mentioning the fact that the scalability problem in edge systems presents significant differences compared with similar computing paradigms, the cloud being the closest one. Indeed, cloud services rely on the large available capacity of data centers and the wide core networks capacity as well as to existing redundancy strategies to provide a set of unlimited and fault-proof services, respectively. Thus, keeping the focus on security aspects, a first strategy to address scalability issues at the edge may boil down to deploying redundant systems. However, the model running at cloud, where any redundancy strategy to be applied leverages a totally controlled cloud environment, does not properly suit what the edge scenario is expected to bring. Indeed, redundancy as a solution does not properly meet the characteristics and needs inherent to edge systems, mainly because the edge scenario is largely diverse, putting together many different stakeholders, what makes control operation by a single institution or organisation unrealistic. Consequently, edge systems may assume control to be handled in a distributed way, thus making the distinct involved stakeholders to participate in the overall control system, beyond being completely disruptive regarding the traditional cloud model, and imposes strong demands to enable the coordination of the different institutions or organisations taking over the control. Moreover, the large diversity of candidate devices to be deployed at the edge makes scalability a bit diffuse. Some questions may arise; for example, does it matter what sort of edge devices scale up? Or in other words, is there any scalability dependence to the kind of edge devices scaling up? Is edge device mobility affecting scalability?

Another important characteristic inherent to edge systems, pretty different from the traditional cloud model and also causing relevant effects on the whole scalability, particularly when dealing with security, is trustworthiness. Indeed, unlike the trustworthy systems belonging to a data center (i.e., owned by the cloud provider), devices at the edge may be owned by third parties, therefore building a large, diverse, heterogeneous, dynamic, and 'non-proprietary' scenario where devices should not be trusted at all. An illustrative scenario may be drafted by a smart car moving around in a city. A non-avoidable effect of such mobility causes the car to get suddenly connected and disconnected to multiple devices while moving, what demands fast security settings. Certainly, relying on the cloud to set a secure connection between a car on the move and a device located close to the car is with no doubt non-efficient in many different aspects, such as scalability or the reduced time where the connection would be up and hence useful. Consequently, to minimise the scalability issues while guaranteeing a good performance, connections settings must be operated at the edge.

Finally, it must be also mentioned that although cloud systems can easily scale up to react to a service demanding more resources (e.g., to guarantee the level of quality demanded by the user), this cannot be widely applied to edge scenarios. Indeed questions related to how edge devices are moving around, how devices may be clustered and shifted among different clusters, potential business models in place limiting wide usability of specific edge devices, how low-capacity constrained devices may be tuned to allocate more resources to a particular service (also extended to rich-capacity devices) just to name a few, remain all yet unsolved.

As a conclusion, it looks at an evidence that security solutions based on centralised cloud architectures cannot be applied at the edge. This is mainly rooted on the fact that because of its centralised nature, these solutions fail to scale when shifted towards the edge, where a huge amount of heterogeneous devices may be found. Consequently, novel scalable solutions must be designed leveraging a decentralised model suiting most edge needs and requirements.

7.1.2.2 Security

Security is a must in any service offered to users and platforms used to manage data or infrastructure. Therefore, security must be a critical by-design aspect in the development of any new technological solution. Aligned to this assertion, it looks at evidence that edge computing demands an efficient security model aimed at sorting out the set of security challenges imposed by a successful and wide deployment of edge computing. Moreover, such a security model must be designed at earlier stages, from the very beginning, as an out-of-the-box feature, so avoiding the need for specific security patches.

However, when moving towards the edge, it must be highlighted that the edge, as a concept, encompasses some well-known characteristics. First, the edge is highly distributed; hence, whatever the solution to provide security guarantees may be, it should follow a completely distributed approach able to scale up as the edge will. Second, the set of devices building the edge is highly heterogeneous (e.g., size, computing and storage, power, etc.) from sensors and actuators to laptops and smartphones; hence, the security solution to be designed must be extendable to support such heterogeneity and devices diversity, as well as the unstoppable evolution edge devices are expected to have in the future, in terms of both new capabilities and deployment volume. Third, edge devices may also be poor devices with limited capabilities, such as small sensors, that fail to meet the requirements to establish secure connections or even to apply any security strategy or solution at all. Although a gateway or a third-party device may be deployed to overcome this limitation, additional security constraints arise as described in next subsection. Fourth, devices at the edge must be authenticated, which is pretty challenging in mobile edge systems. To that end, an edge profile may be used, storing the relevant information for a device (i.e., device specs) to be used by other devices to either be authenticated or selected to execute services at the edge. Certainly, this edge profile should be managed in a distributed way across the edge, otherwise a centralised server will cause scalability problems as mentioned before. Indeed, the whole security architecture must be distributed and must scale as well as the edge will. Therefore, cloud technologies do not suit the expectations brought in by edge systems. Hence, would blockchain be a good approach?

7.1.2.3 Devices computational capacity

The edge scenario, as a whole, is quite diverse and largely heterogeneous, putting together many distinct devices with different characteristics and capabilities. In such a diverse scenario, it seems obvious that several devices will not meet the required characteristics to get endowed with security (e.g., devices with very low computing capacity or highly dynamic devices) or even to establish a direct connection with other systems above in the edge (beyond the system the device is attached to, for example, a wired sensor attached to a small gateway, RaspBerryPi). Therefore, these devices will rely on an additional hardware (i.e., gateway) to send and receive information over the edge. However, this added connection between the devices and the gateway poses a threat to the edge. The northbound connection between the gateway and the systems above at the edge can be secured. However, the southbound connection between the devices (sensors) and the gateway cannot be guaranteed to be secure because these devices are unknown to the edge. Consequently, edge devices receiving data from a gateway should know that such data cannot be guaranteed to be tamper-proof because these data come from less trusted devices or not completely trusted data. However, this scenario may be alleviated if data is contrasted compared with other systems in similar locations. Although this different data comparison may be highly relevant to increase data security, it also introduces the need for a place where to store the collected sensors profiles or information to be used by other devices. Certainly, it must also be said that if this storage is centralised, the problem will come up again. Therefore, the trust issue with non-secure devices and sensors could be addressed by gathering data from multiple devices and sensors and trusting what the majority of the gathered data reports in a specific area. This could be a reasonable procedure to sort out the problem of non-trustable data from unsecured devices.

In conclusion, when a specific device is not trustworthy, adding a gateway for security purposes may not be enough because the device, per se, is not secure and cannot deploy any security policy or strategy either.

A potential trend to sort out this challenge would rely on edge profiles. Profiles may be set for the devices at the edge and shared with the edge systems above, and these may guess what to expect from a gateway. Consequently, should the sensors behind the gateway get attacked, then the expected data (according to the profile) will not match the real received data from the sensor. But would that be enough?

7.1.3 The cloud as a secure architecture 'no longer feasible'

The cloud has become a widely adopted computing paradigm since its design and deployment, mainly rooted on both its resources availability (i.e., high performance) and its resiliency. However, such excellent performance does not mean that the cloud may be the optimal solution for any highly demanding scenario. Indeed, as mentioned, the cloud is a centralised-based model, what undoubtedly brings a substantial impact on some scenarios when claiming for some specific performance needs, such as low latency or security guarantees in services deployed at edge systems.

Indeed, any edge network relying on cloud services to be secure will fail to both scale and deliver the security level that would be expected. For example, let's assume an edge system depends on cloud to be secure. In this scenario, if at any moment, the connectivity to the cloud is lost, what is highly predictable when moving towards the edge, the access to the edge infrastructure by the edge systems will not be permitted because they will be unable to reach the cloud for security provisioning. Certainly, some additional strategy should be designed to fix this scenario, for example, based on distributing the security through the edge.

A key principle behind edge systems refers to the fact that services may run in a completely distributed way throughout the different set of devices deployed at the edge. This paves the way to have devices with no connectivity to the cloud but with access to a network infrastructure at the edge. Therefore, with the security distributed through the edge, a device will be able to request a service from that edge network with no need to reach out to the cloud. Such a distributed security must be deployed by default, distributed over the whole edge network, and thus minimise the need for cloud infrastructure, making the most out of the proximate resources and a better security by design model.

As a summary, theoretically, security may be provided through a centralised model (cloud) or distributed (edge). The first is not properly matching what the edge is demanding, and the second may also be split into two scenarios. The first considers that every edge infrastructure manages and is responsible for its own security, while in the second, the security is completely distributed through the edge, making all elements at the edge participating in a collaborative model.

To build the distributed approach, the current technology best suiting the expected needs is blockchain. Indeed, blockchain offers a totally distributed set of immutable transactions across all participant devices in a network, making devices a part of an 'edge blockchain'. In the scenario, when a device loses connectivity, the local copy of the edge blockchain could be used to authenticate other devices. To that end, the blockchain should be used to store the devices' profiles at the edge, required for service execution and device authentication.

7.2 BLOCKCHAIN IN THE SECURITY ARENA

Blockchain is an easy to deploy technology getting momentum in recent years because of its potential to contribute to solving several existing challenging issues. Indeed, blockchain is suited to sort out challenges coming up when a set of immutable transactions has to be distributed across a network, making it easy for users to verify those transactions. In fact, because there is no central entity responsible for data verification, each user can do it through its local copy of the blockchain. However, as it usually happens with any new technology attracting attention from the scientific and industrial communities, blockchain must not be seen as a universal solution properly addressing any scenario, which is what actually seems to be the case when a novel technology comes up.

In fact, blockchain does suffer from different architectural challenges, out of which scalability and consensus may be highlighted, that may hinder its adoption in some scenarios. On one hand, the scalability issue is motivated by the data stored in the blockchain. Indeed, because the transactions in the blockchain cannot be deleted or removed, the total set of transactions on the blockchain will always grow. To mitigate these scalability problems, particularly when using a blockchain to design a security solution, while simultaneously preventing the overloading of the blockchain with unnecessary data, software developers must seriously consider what may be stored on the transactions and how this information is stored. On the other hand, the consensus algorithm should be carefully chosen, particularly when addressing edge scenarios, because the inherent heterogeneity at the edge puts together devices with different capacities, which may impose highly demanding consensus algorithms not to properly run on constrained devices. In fact, not all consensus algorithms properly meet the edge needs and constraints.

Consequently, with the aim of minimizing the effects of these challenges, it is worth mentioning that blockchain requires a highly efficient utilisation, with a special attention to the storage consumption, because opposite to the cloud, the resources at the edge are pretty limited.

7.2.1 Why to use blockchain

Blockchain is one of the most, if not the most, promising distributed system applications of the last decade. Its ability to keep an immutable set of transactions distributed across a whole network paves the way towards a new world of distributed system applications yet to come. Undoubtedly, this seems to be the proper solution for highly distributed systems.

In practice, when a distributed system must be enriched with security guarantees, different options may be applied. Usual strategies relying on cloud leverage centralised approaches (hence adding a single point of failure) because there is no valid cloud model that can be applied on a distributed system. However, a centralised cloud approach is not suitable for distributed systems, and the solution should be based on moving towards a decentralised approach.

A potential option for distributed security provisioning in highly distributed systems would focus on considering anonymous users, where each user generates its own set of keys, completely anonymous inside the system, requiring the user to manually distribute its public key to the set of destinations to get authenticated. However, this approach is not valid for edge systems where most connections will be machine to machine, requiring automatic authentication of devices. Moreover, edge systems cannot rely on verifying keys manually because this process will not properly scale up.

Another option in the line of decentralised approaches relies on blockchain. Indeed, blockchain is gaining momentum because of its ability to keep an immutable set of transactions, each with capacity to store, for example, the public keys and profiles of the devices. This structure can easily and successfully register a new device into the system, which will be distributed across the edge network by the blockchain. Interestingly, all edge devices will be able to authenticate the newly added device with the known keys and profile.

A key characteristic of using blockchains is the fact that devices do not require a constant network connection because the local copy of the blockchain can be used to validate any device that has already been added into the system. Eventually, if a device loses connectivity, it will be unable to update the blockchain with new devices. However, it will still be capable of authenticating the known devices of its local blockchain copy, and once the device recovers the connectivity, it will be able to update and validate the blockchain to its latest state.

In short, the main rationale and benefits of using blockchain in distributed edge systems is supported by (i) the total distribution of data and responsibilities over the edge; (ii) each node is capable of validating all the devices in the edge network, locally, and therefore, there is no need for any third-party involvement to secure the edge; (iii) edge nodes become able to authenticate other devices even if they are partially disconnected from the edge; and (iv) all devices are empowered with the security of their own data, and they know the edge network and can take decisions based on that information.

7.2.2 The challenges

This section revisits the key challenges that might hinder a wide deployment of blockchain, namely scalability and the data-sharing problem, both of them, as described next, are strongly linked.

7.2.2.1 Blockchain scalability issues

A key and widely known issue in blockchain, certainly hindering its adoption in some particular scenarios, is the scalability. Indeed, although scalability issues may be not so critical in cloud-computing scenarios, where storage capacity is pretty large, these issues become a 'must' when considering the deployment of blockchains in distributed systems, such as the scenario built by edge computing, where memory capacity is usually negligible.

Main scalability issues are motivated by the data immutability characteristic. Indeed, a collateral effect of data immutability propriety is that the data in a blockchain cannot be erased or replaced, therefore leading to an unstoppable growth of the blockchain as new transactions are included. This unlimited growth cannot be eliminated because the only way to do that would be to break the immutability attribute, which is one of the main reasons for using blockchain.

However, although the unlimited growth cannot be eliminated, it seems appropriate to look for initiatives aimed at reducing such a growth. One potential path to that end, would reside on software designers that should be careful when adding data into the blockchain, thus guaranteeing that the data stored in the blockchain is actually essential data, while all non-relevant data is shared using other technologies (i.e., transport layer security [TLS] connections between devices). Another potential solution boils down to deleting the blockchain and setting a new one to reduce space. However, this can only be accomplished if the data stored in the blockchain can be removed; otherwise if such data must be kept, then this solution cannot be implemented.

7.2.2.2 Blockchain data sharing

As it has been stated, no data can be erased from a blockchain once it has been committed. Another collateral effect of such a commitment characteristic refers to data privacy, particularly when sharing private data using the blockchain. This is mainly rooted in the fact that because no one may delete data from the chain, and any device in the chain may access to the data shared, this data may be read by any device. In practice, data is encrypted to not allow others to obtain the data but in case data becomes sensitive and anyone tries to decrypt it. In fact, this is the main reason that makes blockchain unsuitable for data sharing. Instead, data should be shared through either direct channels with other devices or by a third party. In addition, because of the scalability issues mentioned, and to avoid the blockchain to grow under no control at unexpected rates, it is recommended to use the blockchain only for storing static data (i.e., the one that rarely changes or gets updated).

7.3 BLOCKCHAIN ARCHITECTURE

Blockchain is a relatively new technology, whose potential contribution towards security provisioning is currently being explored. Many blockchain implementations exist, some of them linked to various initiatives and projects, such as the Hyperledger project [1]. Beyond the various blockchain implementations, there are even more possible consensus algorithms to choose from, proposing different strategies to achieve the consensus and obtaining different performance and security levels. In fact, a trade-off is set for each algorithm between performance and security.

For example, the proof-of-work (PoW) consensus algorithm applied in Bitcoin ensures the data not to be maliciously changed unless a huge portion of the devices are controlled. However, it consumes a vast amount of resources, making Bitcoin to use more electrical power than some smaller countries in 2019 [2] and what unquestionably becomes a huge performance issue of this peculiar consensus approach. On the opposite side, other algorithms, such as the Redundant Byzantine Fault Tolerant (RBFT), consume much less power but unfortunately tolerate a lower portion of the network being malicious (e.g., PoW 50%, RBFT 33%) [3]. It is also worth mentioning that PoW enables all nodes to participate in the consensus, whereas only a special chosen set of validators can participate in RBFT.

From an architectural perspective, blockchain networks can be classified into three distinct groups, namely permissionless, permissioned, and hybrid or consortium, depending on the consensus algorithm to be used. Interestingly, each group has its own trade-off, and certainly, each group best suits specific environments (i.e., the service or domain where the algorithm is applied), as will be shown in the next section. In addition, two types of consensus algorithms exist, depending on how the validator nodes are defined.

7.3.1 Architectures

This section describes the architectures driven by the specific consensus algorithm in place to be described in Section 7.3.2.

7.3.1.1 Permissionless

In the permissionless blockchain architecture, anyone can join the blockchain network and participate in the consensus because there is no authentication required against other devices when performing consensus operations. Figure 7.2, illustrates the permissionless architecture, being the blue cloud on the blockchain network and the blue nodes the devices belonging to it. As defined in the permissionless architecture, all nodes can participate in the consensus.

This architecture perfectly suits anonymous cryptocurrency networks where participants want to be anonymous to each other and is the basis of the first blockchain ever, that is, Bitcoin. Well-known consensus algorithms belonging to this architectural group used on this kind of blockchain networks are PoW and proof-of-stake (PoS) [4].

Nevertheless, when dealing with security, the fact that anyone can operate on the transactions without any kind of authentication to others brings a serious security issue because a malicious device might perform malicious actions in the network with no one knowing it. This is the reason motivating these architectures to not to be the correct ones when applying security

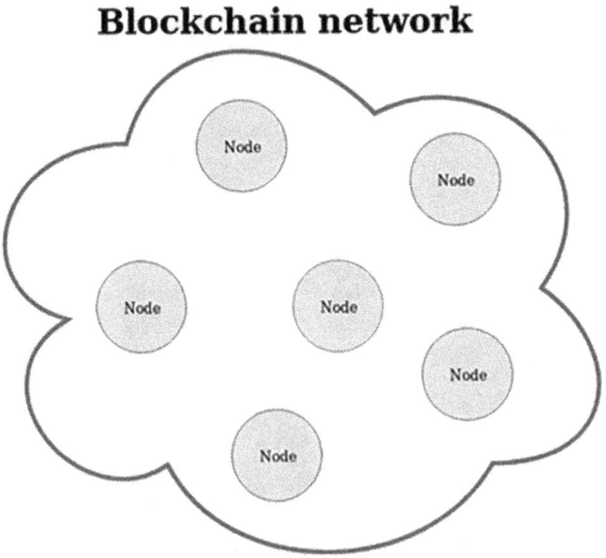

Blockchain network

Figure 7.2 Permissionless blockchain network example.

over an infrastructure using blockchain. In short, if blockchain is deployed to secure a physical edge infrastructure, some trust among edge devices is needed. To that end, the edge provider must control at some level what is being introduced in the blockchain.

7.3.1.2 Permissioned

Unlike the permissionless architecture, in the permissioned architecture, the access to the blockchain network is controlled, and consequently, only authenticated devices can join the blockchain network. Figure 7.3, depicts the permissioned blockchain network highlighting the two main important differences compared with Figure 7.2. First, while similarly to the permissionless architecture shown in Figure 7.2, the blue cloud, representing the network, and the blue nodes, representing the devices belonging to the blockchain network, additional red nodes come up representing the set of nodes outside the private network not enabled to access the blockchain. Second, in this permissioned architecture approach, only the blue validator nodes can participate in the consensus, which is different from Figure 7.2 where all nodes can participate in the consensus.

Usually, in this kind of architecture, consensus is controlled by one single organisation, putting together either one node or a set of nodes to manage all consensus-related tasks. This makes this architecture a non-optimal choice for some scenarios (e.g., bringing together a large set of distributed nodes and multiple networks), rooted in the fact that the blockchain, by

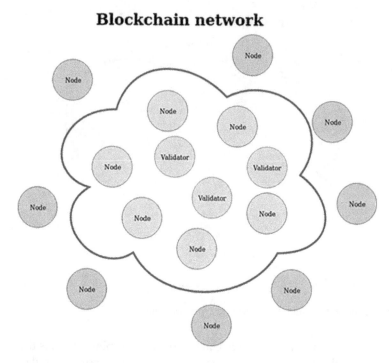

Figure 7.3 Permissioned blockchain network example.

definition, should be at least public to be read and free to join from any device. In fact, this architecture trespasses the boundaries of what a blockchain should be, mainly because security cannot be guaranteed, since the consensus is controlled by a single institution and is more prone to failures. In addition, imposing such control on the blockchain membership goes against the blockchain concept itself.

7.3.1.3 Hybrid

In this architectural approach, any device is allowed to read the blockchain, that is, any device can verify its data, but only a controlled set of 'validators' is allowed to perform consensus operations and commit blocks. Similarly to the two previous architectures, Figure 7.4 represents the blockchain network as the blue cloud and the devices belonging to it as blue nodes. As in Figure 7.2 (i.e., permissionless architecture), any device can join the network, but differently, not all devices can participate in the consensus. Indeed, two types of devices may be defined, normal devices (belonging to the blockchain network but cannot participate in the consensus) and validators (can participate in the consensus).

Blockchain network

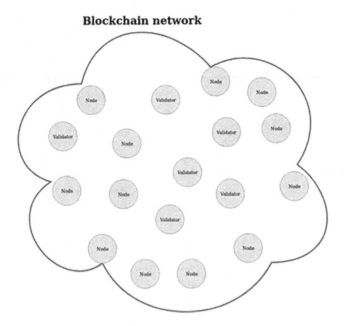

Figure 7.4 Hybrid blockchain network example.

Acting as a hybrid approach between the two other approaches, it seems a reasonable solution to provide security with no need to lose the complete control. Indeed, this approach enables all network devices to review and verify the blockchain data, while simultaneously enabling a consortium of organisations to control the blockchain through a set of validators owned by the consortium members. By doing so, the blockchain cannot be altered by any attacker as long as the validators are secure.

In short, the benefits brought in by the hybrid architectural approach to provide security may be summarised as (i) its capacity to deploy a controlled consensus that can establish trust to all devices; (ii) the capacity to handle the consensus through various organisations forming a consortium, thus making the system more robust; and (iii) the capacity for everyone to read the blockchain with no need for an authentication process.

7.3.2 Consensus

In the previous section, the different blockchain network architectures that may be designed, highly linked to the different consensus algorithms to be deployed, have been described. This section goes deep into these consensus algorithms.

Several consensus algorithms are being used in different environments and domains. A key component characterizing these algorithms refer to the set of devices enabled to run the consensus and can be characterised into two groups, public and permissioned. On one hand, public consensus algorithms enable any device to participate in the consensus. However, in this strategy some issues may arise, mainly motivated by the fact that because all devices can perform operations over the blockchain, the security of the blockchain will be left at the devices will. Thus, should a device decide to perform a malicious activity, it will be hard to detect by other devices. On the other hand, permissioned consensus algorithms leverage a controlled set of validators to assure the authority over a blockchain network. A key characteristic for permissioned consensus algorithms is its scalability, motivated by the fact that only a well-determined set of nodes handles consensus tasks, which indeed, turns into a much better performance, and makes permissioned consensus algorithms to seem the proper ones when applying security over a blockchain network.

In short, in a permissioned consensus algorithm, only devices appointed as validators may vote for new blocks and transactions to be committed into the blockchain. Consequently, the consensus is totally controlled by the set of validator devices, while normal devices only observe and receive the validators votes. Finally, when devices have enough votes for a block from the validators, the block is considered to be committed.

Moreover, when accepting a new block, the set of validators can also check some other information regarding the block. At the end of this chapter, an overview of potential considerations to be taken into account when adding new transactions is included. Just as an illustrative example, a transaction performing a privileged operation can be scaled up to another validator, device, or cloud to perform a privileges check.

7.3.2.1 Public algorithms

Some widely known public algorithms are:

PoW: Proof of Work [4]
> In the PoW consensus algorithms the miners concept is defined. 'Miners' are the peers creating or mining new blocks to be added to the blockchain. For these blocks to be accepted by others, they must meet a special hash starting with four consecutive zeros. Thus, miners must try different combinations of transactions and calculate multiple hashes of a block until they find the required pattern. Once the pattern is found, the block is distributed by others and committed. This peculiar consensus algorithm requires a high amount of computational power – this consensus algorithm is using more power than some small countries in 2018 [2], which certainly makes it not to be the proper one for edge scenarios.

PoS: Proof of Stake [4]

This consensus algorithm chooses the validators following a stake system. The stake system favours the users with more wealth (stake). Therefore, a user can only behave maliciously on the blockchain if it owns 51% of the total stake of the system. This algorithm follows the criteria that the users with most stake on the system are the ones with more to lose. Therefore, they will complete the transactions correctly when chosen as validators.

7.3.2.2 Permissioned algorithms

Some widely known permissioned algorithms are:

PBFT: Practical Byzantine Fault Tolerant [5]

PBFT is the most basic Byzantine fault tolerant algorithm. It relies on distributed rounds, each adding a new block. The process starts by first selecting a primary (validator) out of the set of possible validators. To that end, some policies are applied, customised to the final objective and environment of the protocol. Once the primary is selected, it becomes responsible for setting the order for the different transactions as they are being received by the primary. Then, when the primary receives a petition to add a new block, the primary runs the commit protocol, which is structured into three different phases, pre-prepared, prepared, and committed, moving a necessary condition for a node to pass each phase to receive votes from two-thirds of the validator nodes. Certainly, the PBFT protocol requires every node to be known by the blockchain network.

RBFT: Redundant Byzantine Fault Tolerant [6]

The RBFT algorithm extends the previous PBFT algorithm, intended to addressing the performance problem coming up when the faulty or malicious replica is in fact the so-called primary. To that end, RBFT makes all clients to a primary and introduces the concept of the 'master instance', that is, all clients and primary instances are responsible to order the requests but only the master instance order is executed.

The authors in [6] present some promising results for RFBT, showing the performance when no failure occurs to behave similarly to other protocols, while the loss of performance is only 3% compared to the 78% for other protocols when a failure comes up.

ZYZZVA: Speculative Byzantine Fault Tolerant [7]

Also inferred from the Byzantine fault tolerant algorithm, this algorithm introduces the use of speculation to both minimise the cost and simplify the design compared with traditional Byzantine fault tolerant algorithms. The validators respond automatically to any request, skipping the three phases commit protocol proposed in PBFT. Therefore, they follow the order of the so called 'primary'; thus, because the three-phase commit protocol is never executed, the network communications are reduced to theoretical minimums. However, some inconsistencies may come up that must be addressed by the validators and devices.

7.3.2.3 New transaction acceptance

As opposed to the public blockchain algorithms or other permissioned consensus algorithms, there are extra factors that may be considered when accepting new transactions and thus creating new blocks, particularly when facing highly constrained and heterogeneous scenarios. Let's assume an edge scenario, where different edge devices should be endowed with security guarantees. In this scenario, an edge device must not blindly accept any transaction because this would make the system insecure. A solution to handle this issue would require this device to check for each transaction, what consequently requires a centralised service performing the checks, driving again towards a centralised solution, which is not good at all. In fact, the edge must follow a distributed system architecture, while at the same time providing security. Aligned to that statement, edge systems must be careful around the policies applied when adding a new transaction to check both the new transaction does not to violate any security requirement and the issuer of the transaction has the right to perform the operation. Ideally, this information should be stored in the blockchain, too, so any lookup becomes a local query, making the edge really distributed with no need to rely on the cloud or any other centralised agent. However, when this is not the case and a third party should be also considered, the requests should be kept to a minimum, only demanding a really small set of operations, aimed at minimising the centralised dependence of such request.

7.4 BLOCKCHAIN SERVICES IN EDGE COMPUTING

This section introduces several possible solutions aimed at providing security to devices at the edge. All solutions described in this section leverage blockchain technologies, considering a blockchain putting together all devices in the network, and each device associated with a profile, including the device public key. The section starts by presenting a method to create a secure device profiling over the blockchain. Afterwards, a method for authentication using the blockchain device profiles is presented. Finally, a method for sharing signed data with the blockchain using the previously described profiles is shown.

   Figure 7.5 presents the most basic blockchain edge network among those described in this chapter. Indeed, devices and profiles are included inside the blockchain network, and each device has a copy of the data (all the other profiles). Consequently, each device will be able to locally verify any

Blockchain network

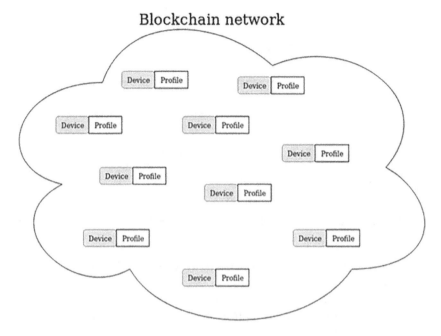

Figure 7.5 Blockchain network example with profiles.

other device through the blockchain using the profiles, with no need to rely on any other entity or third party.

7.4.1 Secure device profiling

Providing security at the edge requires all edge devices to be identified and secured. To that end, the edge, as a system, requires an element responsible for helping other devices identify the components in the edge network and access its profiles. This is certainly critical because the profiles are stored in the cloud, and this is largely known not to be feasible or scalable for edge scenarios. Therefore, the edge requires a technology that allows us to store those profiles in a distributed way, over the edge, and this technology may be blockchain.

Blockchain provides us with the distribution of data and the immutability of that data. Following these properties, the edge can distribute a device profile using the blockchain in a secure manner. The profile can be stored in the blockchain inside a transaction, and it can be exposed as a device ID (see Figure 7.6). With this data structure, the blockchain will store the session state of the edge devices, this session would be composed of device identifiers and data, and it would maintain all blockchain properties.

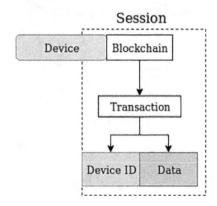

Figure 7.6 Session example with devices profiles in form of transactions on the blockchain.

To offer trust on the blockchain profiles, the transaction must be verified and partially controlled, so no malicious device can introduce malicious profiles. Therefore, it seems that a permissioned consensus algorithm in a hybrid architecture may be the proper option for a successful blockchain deployment at the edge.

Once the transactions and profiles are trusted, the device IDs are required to be trusted. To do that, the profiles will require a key pair. Elliptic curve cryptography (ECC) will perform better than RSA because the former has better performance with less key size. Then, the private key of the pair will be held by the device and the public key will be stored on the blockchain. With this strategy the public key will become the device ID, and the profile will only be verified to the device owning the private key.

Although this storage by itself does not do much in terms of security because the transactions and blocks have to be approved by the validators, the edge can actually use the blockchain to perform a public key distribution of approved devices at the edge. This will act as the edge session that will tell other elements that a given key belongs to the edge and should be trusted, as seen in Figure 7.7.

But again, this structure alone does not provide any flexibility at all because the public key cannot be associated with anything, which is also motivates that every transaction should have a rule-set. Indeed, a rule-set is defined as a set of actions that are applied to the session for a given ID (public key). For example, when a device registers into the session providing its public key, a rule-set of registration will be added. Afterwards, this profile may be used, for example, when choosing devices to execute edge services or simply when performing machine-to-machine connections, so a device can know if is connected to another device.

When performing a registration, the device to be registered will provide its edge profile, including the device capabilities and sensors. The capabilities

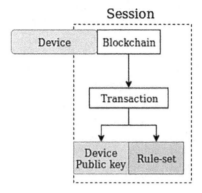

Figure 7.7 Session example with devices profiles in form of transactions on the blockchain, containing a public key and a rule-set.

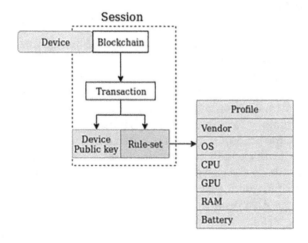

Figure 7.8 Session example with devices profiles in form of transactions on the blockchain, containing a public key and a rule-set; the rule-set contains the specific edge profile.

are defined as what a device is, for example, as shown in Figure 7.8: Vendor, OS, CPU, RAM, GPU, battery, or AC power. Certainly, this is just an example, and many other different characteristics and attributes may be included in the edge profile. These capabilities will be used to select the set of resources best suiting what a particular service demands to be successfully executed. Hence, in short, the profile will keep a register of the device attached sensors and will help other devices to know what the device sensors are and what they can provide.

Assuming that the registration process is done correctly and the device registers with a real profile, then that device will be secured into the session blockchain for the lifetime of the blockchain. In fact, even assuming an attacker gains access to a device and its private keys, it will never be able to modify the registered profile because the blockchain is immutable by definition.

Another thing that can be done with this structure is the registration of a second public key; this public key will be an ED25519, and it will be used by the devices to sign data and later send it to any session device. The addition of this signing key enables the devices in the session to send unencrypted signed data, what is mostly intended for sensors. Most of the edge sensors will be placed on public places, and it makes no sense to hide that data when the sensor objective is certainly to share that data with the edge. Consequently, the sensor can use a signing key and the edge session to sign the data, then send the signed data (over the desired network connection or third channel), and finally the receiver device verifies the signature and authenticates the sending sensor against the edge session.

7.4.2 Authentication

Edge computing as a concept, leverages a large amount of devices usually connected among themselves, through machine-to-machine communications, to share data and resources. This scenario, with a large amount of devices, heterogeneity, and mobility (devices suddenly getting in and out) requires an efficient authentication strategy particularly tailored to face these edge-system constraints. As stated r, blockchain can become a really powerful technology to perform this authentication between edge devices.

To that end, the procedure to set a secure connection between two devices belonging to the blockchain network or the session is as follows. First, each device can compare the public key of the TLS connection with the public key of the session, and if that public key belongs to the session, then it will have a rule-set with a profile associated with it as seen in Figure 7.9.

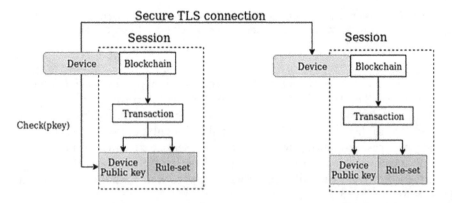

Figure 7.9 Device establishing TLS to a second device. The device can verify the identity of the device using the profile. TLS, transport layer security.

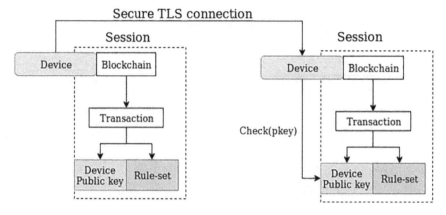

Figure 7.10 Device establishing TLS to a second device. The second device can verify the TLS key against the session. TLS, transport layer security.

In a second step, the rule-set profile will be checked to verify and know who the incoming connection is or who is the one willing to connect, as seen in Figure 7.10.

The key benefit of this approach is that the validation is performed locally and totally distributed between session devices with no need to connect to any centralised entity to verify the data, thus becoming a highly scalable solution.

7.4.3 Integrity of the data provided by the edge

There are two major fields when talking about the integrity of the data at the edge. The first refers to the need for validating any data sent by any edge device. The second deals with the validation of the data sent by edge devices behind gateways and, therefore, unknown to the edge.

7.4.3.1 Validation

The data validation procedure in a device may run two approaches:

- A device can validate the data received by verifying the TLS connection. In this case, because the TLS is guaranteed to be secure and the public key of the TLS connection can be verified against the one in the session profile, the data is guaranteed to be correct (see Figure 7.10).
- Alternatively, a device can send signed data, but it is unencrypted (i.e., a sensor broadcasting a temperature sensor data to the network). Then, any device receiving the data will be able to verify the data sender by verifying the data signature using the session, as seen in Figure 7.11.

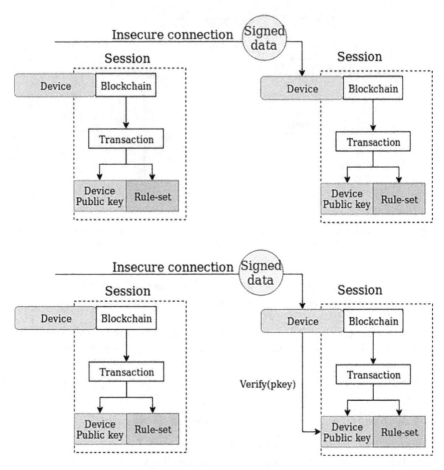

Figure 7.11 Device sending signed data over the edge, the receiving device can verify the data.

7.4.3.2 *Gateways*

Using the previously stated rule-set, the edge can also register a different type of device, which is a device that at the same time has various devices, referred to as a 'gateway'. In edge computing, most sensors are behind gateways that provide access for and to them. Thus, those sensors are unable to be reached from the network or, in other words, are unable to reach outside the gateway (see Figure 7.12). For this reason, the gateway can be registered, considering the different devices it has attached to, with a new rule-set. This will tell other edge devices that the devices and sensors exist behind a gateway, but that those devices can be reached through the gateway only – although in the case that at the same time the

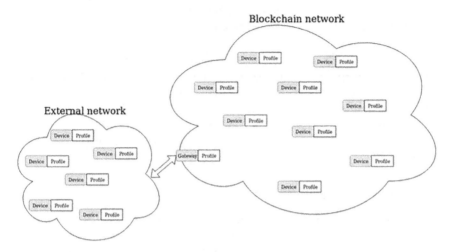

Figure 7.12 Representation of an edge network with a gateway.

gateway registers the keys of these devices with the rule-set. Therefore, all edge session participants will be able to authenticate the data received from the devices behind the gateway.

The data provided by a gateway will be verified using the same method used by any device. As explained before, because the only change is the actual session information – letting the device know whether the data comes either from a device or a gateway – the verification will still be performed with the public key extracted from the profile of the gateway.

The session representation for a gateway will be as seen in Figure 7.13. The gateway is a device that at the same time can register subdevices, with its own keys and profiles. Thus, when registering on the session, the gateway registers its subdevices at the same time.

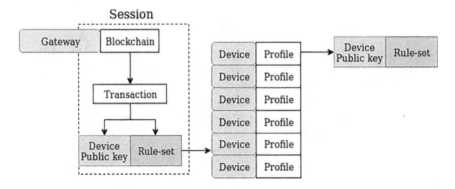

Figure 7.13 Representation of the gateway with the rule-set and the subdevices profiles.

7.5 BLOCKCHAIN VERSUS CLOUD SOLUTIONS APPLIED TO EDGE SCENARIOS

This section briefly summarises blockchain compared with other cloud solutions when applied to edge systems. First, a review of the authentication model of CAs and certificates versus the authentication model with blockchain is presented. Second, a review of the device profile with centralised cloud servers versus the blockchain is also introduced. Finally, the integrity of the data at the edge, comparing databases and blockchain approaches is discussed.

7.5.1 Authentication model

When comparing blockchain as an authentication model with the widely used CAs and certificates, it must be pointed out that certificates are designed to be static, thus becoming the best option when dealing with centralised security, for example, for centralised web services. However, CAs and certificates are difficult to apply when dealing with the edge, putting together a set of heterogeneous and distributed mobile devices.

Moreover, blockchain is completely distributed, hence radically changing the way CA and certificates work. Indeed, when using blockchain the single point of failure of the centralised servers is removed, while at the same time it adds the fact that data can be constantly distributed to peers through the blockchain. This makes the blockchain a perfect key distribution system, highly available and decentralised, while being highly customisable, because any data can be stored as a transaction as long as it makes sense to the edge and follows the good practices. In fact, in an edge scenario, blockchain can be used to authenticate devices between them, using profiles stored in the blockchain, so all edge devices may know who the other devices at the edge are and, consequently, can authenticate them using the local copy of those profiles, even if they are only partially connected to the edge.

7.5.2 Secure edge device profiling model

To keep the profiles for edge devices, if the edge uses a centralised database or even a distributed database, it will soon run into scalability problems. This is because the amount of devices will overpass the database capacity. Additionally, there is a limit of replication for a database because the more replicated it is, the more time it will take for the whole database to be coherent on all its instances.

A key characteristic of the blockchain technology relies on the fact that it offers a completely distributed set of transactions. Consequently, if the whole edge system stores edge profiles as well as all relevant edge information on a blockchain then, there is no central entity needed for any action. This is undoubtedly a notable add-on because if the edge would require a central entity to identify any edge device with its profile, then a huge bottleneck and a poor scalability structure will result. Moreover, in this case, an active connection to cloud will always be required, which is not desirable either.

On the other hand, if the edge relies on blockchain technologies, then as has been stated, the information of the profile will be replicated on all the devices. Thus, any device can access that information locally, even without connection to the cloud, and can actually access device profiles while being offline and use them even if the other device is the only one reachable on the network.

Indeed, the main benefit of deploying blockchain at the edge when talking about cybersecurity may be synthesised as no need for a centralised entity verifying profiles, but instead the edge can do that same operation on the blockchain in a distributed way, which does away with scalability as a problem.

7.5.3 Integrity of the data provided by the edge model

Using a centralised database at the edge drives all data to be stored in one single device. Thus, if such device happens to fail, so will the data. Even more, as already stated, such a centralised system will fail to scale up with the edge devices. Therefore, it will be impossible for one single device to serve the whole edge. Indeed, although the centralised database can be kept at the cloud, it will still suffer from all the issues mentioned. Consequently, centralising the information that the edge must have access is a bad choice. In addition, centralizing the data in a single point brings significant security risks because the data can be tampered with if an attacker gains access to the database. Similarly, in a distributed database scenario, the data may become fail-proof, but the scalability becomes an issue when using such database in a highly distributed network, as the edge is.

On the other side, blockchain can totally distribute the data over the edge devices and can scale as good as the edge will. This is really important when distributing data securely because a database can become a single target of an attack, while the data in a blockchain, distributed across the edge, is nearly impossible to be tampered with because of the replication level and immutability properties.

7.6 CONCLUSIONS

Blockchain seems to be a powerful and ideal tool to secure large distributed systems, such as the envisioned systems to be considered in edge computing. In fact, blockchain offers the possibility of distributing and maintaining an immutable data set across all edge devices. This is much more powerful and suitable than any potential cloud strategy suffering from network scalability issues and single-point-of-error problems when deployed at the edge.

However, blockchain technology comes at the price of more resources utilisation at the edge and must also address, as explained in Chapter 2, various architectural issues, such as the scalability of the blockchain-stored data.

Therefore, assuming blockchain to be successfully implemented at the edge, then the distributed data can be used to secure edge systems, mainly motivated by its immutability property, as proposed in Section 7.4. Indeed, blockchain data can be used to establish a chain of trust between validators and edge devices and to distribute the device profiles all over the edge. Later on, those profiles can be used to establish secure connections and to authenticate the devices against other devices at the edge, without the need for any connection to the cloud or any third party required to validate the device profiles.

Finally, we may conclude that blockchain benefits to guarantee security provisioning for edge systems would keep growing as long as research efforts devise innovative solutions to mitigate current drawbacks and limitations.

ACKNOWLEDGMENT

This work was partially supported by the H2020 EU mF2C Project ref. 730929 and by the Spanish Ministry of Economy and Competitiveness and by the European Regional Development Fund, under contract RTI2018-094532-B-I00 (MINECO/FEDER).

REFERENCES

1. An Introduction to Hyperledger. Available online at 'https://www.hyperledger.org/wp-content/uploads/2018/08/HL_Whitepaper_IntroductiontoHyperledger.pdf' Date: April 29, 2019.
2. Vranken, Harald. 2017. Sustainability of bitcoin and blockchains. *Curr. Opin. Env. Sust.* 28: 1–9.
3. Z. Zheng, S. Xie, H. Dai, and H. Wang. 2017. An overview of blockchain technology: Architecture consensus and future trends. *Proceedings of the 2017 IEEE International Congress on Big Data (BigData Congress)*, June 2017, Honolulu, HI.
4. Torre Dominique and Sothearath Seang. 2018. Proof of work and proof of stake consensus protocols: A blockchain application for local complementary currencies. *Proceedings of the International Symposium on Money, Banking and Finance, Sciences Po Aix.* June 2018, Aix-en-Provence, France.

5. Miguel Castro and Barbara Liskov. 1999. Practical Byzantine fault tolerance. In *Proceedings of the third symposium on Operating systems design and implementation (OSDI '99)*. USENIX Association, Berkeley, CA, pp. 173–186.
6. Ramakrishna Kotla, Lorenzo Alvisi, Mike Dahlin, Allen Clement, and Edmund Wong. 2010. Zyzzyva: Speculative Byzantine fault tolerance. *ACM Trans. Comput. Syst.* 27, 4, Article 7 (January 2010), 39 p. http://dx.doi.org/10.1145/1658357.1658358
7. P.-L. Aublin, S. B. Mokhtar, and V. Quema. 2013. Rbft: Redundant Byzantine fault tolerance. *Proceedings of the 2013 IEEE 33rd International Conference on Distributed Computing Systems (ICDCS)*, July 2013, Philadelphia, PA.

Chapter 8

Blockchain based solutions for achieving secure storage in fog computing

Ouns Bouachir, Rima Grati,
Moayad Aloqaily, and Adel Ben Mnaouer

CONTENTS

8.1 INTRODUCTION

Technological advances have covered all the sectors resulting in a world of huge number of connected devices. We have moved from a simple call between two users to the Internet, then to the Internet of Things (IoT), to the Internet of Everything (IoE), and recently, to the concept of edge and fog computing [1,2]. We are surrounded by a significant number of devices and gadgets (such as mobile phones, laptops, vehicles, watches, cameras, sensors, smart home appliances, etc.) that are becoming essentials because they facilitate and simplify our lives. Everything can be done through the network: a simple text message; financial transactions and also home security or health monitoring, and the list is quite long. For instance, it is possible to track a patient's health situation in real time independently of location. Also, it is possible to monitor the security or other things like the environment condition of homes from distant locations and to make financial transactions anytime. This is possible by using smart devices and gadgets that able to collect data about these transactions, or sensed information, and to send it through the network to the cloud servers where it is analysed and stored to be reused when needed. In other words, we have a substantial number of mobile networked objects, interconnected through the ubiquitous network and continuously producing and exchanging data that can be used in several applications and sectors such as communication, vehicular, health care, environmental, etc. Traditionally, the collected data is stored in cloud servers basic databases following a centralised approach [3,4]. However, the explosion in the volume of data produced by smart devices makes these applications face many challenges and the following questions arise:

- Where to store this big amount of data?
- How to efficiently compute and extract this data when needed?
- How to guarantee the security, integrity, and privacy of this data?

Indeed, the collected data is sent to a remote cloud to be processed, analysed, and stored; however, the exiting centralised cloud system is difficult to scale with the amount of the generated data. Efficient and sufficient computing resources are also required to execute substantial applications by providing the needed information with minimum delay. Moreover, this data may contain several types of information that can be sensitive and confidential and should be protected. Thus, securing this system from malicious hackers trying to send wrong data or to get access to stored information is an important challenge.

The best way to overcome these challenges and provide the highest system efficiency, modern technologies and paradigms should be included such as edge and fog computing and blockchain.

Fog and edge computing is a new distributed data storage structure that cope with the traditional cloud issues. It consists of many interconnected computing sources located at the edge of the networks considered a single logical entity, capable of performing distributed computing and storing services, as it is on a single device [5].

Blockchain technology is defined as the peer-to-peer (P2P) distributed ledger used to record approved events and transactions. In other words, it can be described as distributed database containing all approved and shared data among all validated users. No data should be added to the blockchain database or interrupted without verification. BC is a promising solution to provide trust, security, and privacy in many cryptocurrency systems including communication networks [6]. Blockchain is also being used to manage heterogeneous IoTs. The authors in [7] have proposed an architecture for blockchain IoT, which keeps track of data exchanges between consortium network participants.

This chapter focuses on the integration of IoT systems with fog and edge computing structure and blockchain paradigm because the combination of these three technologies leverages the benefits of each one. It is organised as follows: Section 8.2 presents an overview of IoT, fog and edge computing, and blockchain. The challenges facing each of these technologies are detailed in Section 8.3. Then, Section 8.4 discusses the advantages and issues of this integration. Some solutions provided in the literature in several sectors are presented in Section 8.5. Finally, Section 8.6 concludes this chapter.

8.2 BACKGROUND

Three of the most popular technologies are IoT, blockchain, and edge and fog computing. They provide promising solutions to enhance and facilitate our lives by allowing many technological applications to be created. This section presents an overview of these three paradigms

8.2.1 IoT overview

Recent technological advances in electronic and wireless communications have introduced a revolution to the IoT by including tiny devices and allowing them to contribute to the collection and exchange of data. Indeed, these advantages have allowed small-size, cost-effective, low-power, and multi-functional sensing systems to be built that enable communication, monitoring of several phenomenon of concerns, and follow trends over time. Such systems are used today in many fields like smart homes, cities, vehicles, e-health, agriculture, and the lists continue. For instance, it is possible to monitor the patient's health remotely by gathering physiological health information in real time in a more convenient and less-intrusive way, which helps in the early detection of potentially abnormal health situations. Moreover, thanks to several gadgets that can be added to homes, it is possible to control its security anytime and anywhere (i.e., video surveillance) or to monitor home devices (such as fridge, AC, TV, etc.) [8]. Driven from IoT systems, other domains have emerged such as the Internet of Vehicles (IoV), Internet of Energy (IoEnergy), Internet of blockchain (IoBlockchain), and Internet of healthcare (IoHealthcare) making life easier by using ambient intelligence, assisted driving, new services, and advanced applications [9,10].

8.2.2 Blockchain overview

Blockchain is a tamper-proof, encrypted, distributed ledger that does not have a central party that establishes trust. A blockchain consists of a linked list of blocks that contain information, referred as a 'set of transactions'. Each chain of blocks is distributed; this means that the blockchains are decentralised across a network of blockchain participants. Each block is cryptographically chained to the previous one by including its hash value and a cryptographic signature. Participants on the blockchain verify all transactions using consensus mechanisms, and that is how trust is guaranteed (Figures 8.1 and 8.2). The trust is monitored by the open source code backed by cryptography. In proof-of-work (PoW) blockchain (type of consensus mechanisms), each new transaction has to be verified by all participants in the network such as Bitcoin and Ethereum. Using the other type of consensus mechanisms, proof-of-stake (PoS), the verification is performed by a selected participants in the network. These consensus mechanisms make the blockchain secure and tamper-proof.

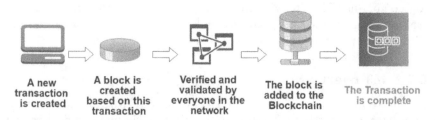

| A new transaction is created | A block is created based on this transaction | Verified and validated by everyone in the network | The block is added to the Blockchain | The Transaction is complete |

Figure 8.1 Blockchain mechanism.

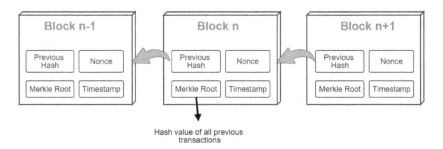

Figure 8.2 Blockchain structure.

8.2.2.1 Blockchain versions

The architecture of blockchain was, from the beginning, designed to handle big variety of implementations. Indeed, the three generations of blockchain namely Blockchain 1.0, Blockchain 2.0 and Blockchain 3.0 could revolutionise many fields. Blockchain 1.0 includes cryptocurrencies, financial transactions, or in general, decentralisation of money and payments. Blockchain 2.0 and Blockchain 3.0, providing an enhanced usage of the blockchain, have emerged almost in parallel in an explosive manner around 2015. Blockchain 2.0 is commonly used for digital finance, and Blockchain 3.0 is used for digital society. The main idea of Blockchain 2.0 is exchanging values in a distributed manner, and unlike Blockchain 1.0, in which the value being transferred on the network is in the form of a currency, Blockchain 2.0 transfers programmable transactions in the form of smart contracts defined as programs and rules that self-execute when certain conditions are met. A smart contract is a transaction embedded to blockchain that creates a powerful and decentralised network of computing and contains more enhanced logic. It stores its own data and can access other resources to analyse and evaluate its current state and perform actions. The original intention of blockchain was to create a public network, which is referred to the model of Bitcoin, Ethereum, and Litecoin, but, because of some limitations (its complete openness and high cost in terms of money, time, and energy to process transactions), private blockchains have been developed as well.

8.2.2.2 Public and private blockchains

The difference between public and private blockchain is based on monitoring who can participate in the network and in which parts and who can manage the ledger. A private blockchain is considered closed network. To join it, an invitation is required. Private blockchains are faster and the transactions cost is lower because of the reduced number of processing nodes. *Although*, they are more complex in term of the process to join the blockchain because of the control access. There are also hybrid solutions that combine private and public blockchains referred to as 'consortium blockchains' [11].

8.2.3 Fog and edge overview

With the explosion of the number of connected devices and amount of exchanged information, it is becoming more and more difficult for the cloud architecture to handle all this data. The decentralised approach can create delays and performance issues for connected devices. To design a high-performance architecture in scalable networks, a decentralised data storage system was proposed using edge and fog computing.

Fog and edge are more recent architectures that allow the data processing from the centralised cloud to move to the edge of the network. Indeed, because it is expected that the number of connected devices will keep increasing, the amount of communications to be handled by a cloud will grow remarkably. Moreover, some applications are latency-sensitive; they require a fast response and mobility support, such as health care and traffic light system in smart transportation. For these applications, delays caused by transferring data are unacceptable because they may cause many dangerous problems. Therefore, the cloud model capacity should to be enhanced [3]. Edge and fog computing have been used to cope with this challenge to achieve better quality of service and experience by pushing more intelligent computing resources, networking, and storage capacities closer to the devices, and providing various benefits such as handling huge amount of data and faster response. Table 8.1 summarises the differences between cloud and fog and edge computing.

Table 8.1 Comparison of cloud and fog computing

	Cloud computing	*Fog computing*
Geographical distribution	Centralised	Distributed
Location of the service	Within the Internet	At the edge of the network
Distance to end devices	Through multiple hops (far)	Single to few hops (near)
Number of server nodes	Few	Many
Communication system	IP network	WLAN, 3G, 4G, IP
Storage capacity	High	Multiple
Latency	High	Low
Mobility	Limited	Supported

Abbreviations: IP, Internet protocol; WLAN, wireless local area network.

8.2.3.1 Fog versus edge computing

It seems for many that the terms 'edge' and 'fog' are similar because they share several keys of similarities. Both fog computing and edge computing provide the same characteristic of pushing data processing and analytics close to where the data originated from (i.e., sensors, watches, motors, screens, speakers, etc.). Edge tends to process the data locally on the device or sensor itself, without being transferred anywhere or on the closest gateway device to the sensors. The data is stored locally, ignoring the aspect of integration. Edge maintains the data on the device were it is generated, which keeps it discrete. However, the small processor and storage capacity of these devices can be saturated quickly with fog computing, data is processed by fog nodes, connected to the local area network (LAN) or into the LAN hardware itself. It allows data gathered from several devices to be handled at once, which improves edge's capabilities by processing real-time requests efficiently. Fog and edge maybe combined. In other words, fog computing can be considered the standard that defines how to use edge computing. It facilitates computing and storage operations and the network services between the devices and the cloud (end-to-end architecture).

8.2.3.2 Fog computing structure

Fog computing is a promising computing model that enhances and brings computing resources to the edge of networks [3,5]. As distributed infrastructure, it involves a set of high-performance physical machines, called 'fog nodes', linked to each other and considered as a single logical entity. These fog nodes reside at the edge of the network, close to the source of data, and can locally collect, analyse, and classify the data rather than sending all data stream to the cloud. This results in important traffic mitigation in the core and faster processing of big data services to perfectly achieve the vision of systems such as IoT. These nodes can interact, when required, with the cloud (e.g., for long-term storage). Figure 8.3 shows the architecture of network-based fog computing. In this architecture, fog computing can be considered a layered service structure and an extension of the cloud-computing paradigm that provides more localised real-time monitoring and optimisation for IoT applications, while the Cloud provides global monitoring, optimisation, and other advanced services.

Fog architecture consists of three layers:

- *Terminal Layer*: It includes end-user devices, mobile devices, and IoT devices. These devices are in charge of generating the data by sensing objects and events and by sending this sensed data to the upper layers for processing and storage.
- *Fog Layer*: It is the middle layer, located at the edge of the network and widely geographically distributed. It includes large number of fog nodes such as routers, access points, servers, switches, etc. They are all linked

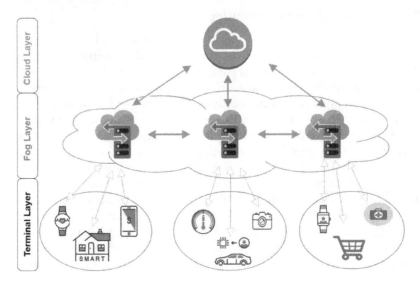

Figure 8.3 Fog computing architecture.

together, and they collaborate to store, compute, transmit, and receive sensed data. Fog nodes communicate with end devices mainly by wireless access technologies, including WiFi, 3G, 4G, Bluetooth, and so on, and each fog node is connected with the cloud through an Internet protocol (IP) network (Internet). Fog nodes analyse and store the data generated by terminal layer devices, and the only important and valuable data is forwarded to the cloud layer for storage or next processing.

- *Cloud Layer*: It includes powerful computing and storage capabilities able to handle extensive computation analysis and permanently store the enormous amount of data. For efficient use of cloud resources, not all computing and storage tasks go through this layer.

The biggest characteristic of fog computing, and the most significant advantage compared to the other traditional computing systems, is that it provides the cloud the capability close to the end user. Data is processed locally by the widely geographic distributed fog nodes, which enhances the data transfer delay and the network transmissions.

The fog computing paradigm will efficiently meet the demands of real-time and low-latency data analysis and decision making.

8.3 EMERGING ISSUES AND CHALLENGES

The use of IoT, fog and edge computing, and blockchain may provide many advantages to our lives and data; however, these technologies may face many issues and challenges.

8.3.1   IoT general challenges

Typical IoT applications are based on wireless sensor networks (WSN) communications to sense and transfer data to a storage data centre where it can be processed, analysed, and stored. The important technological advances have led to an expansion of the IoT applications that will result in a large number of mobile-connected objects, continuously generating and exchanging data, deployed across large geographic areas, and mutually interconnected through the ubiquitous IoT. The consequent explosion in the volume of data produced by smart devices has made the IoT system face many challenges [5,12] (Figure 8.4):

8.3.1.1   Quality of Service (QoS)

IoT applications include various type of data such as emergency response, augmented reality, video surveillance (real-time traffic), speech recognition, computer vision, and self-driving. They have different requirements in term of QoS such as delay, throughput, and reliability that should be respected by the IoT network. In highly scalable system, a large number of devices with various requirements in term of QoS that may change with time are sharing the same networking resources. The IoT network should be aware of this variety and able to provide each device the expected service to guarantee high level of application performance.

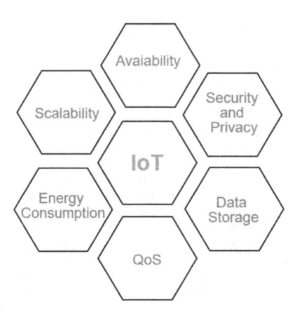

Figure 8.4 Internet of Things (IoT) challenges.

8.3.1.2 Security and privacy

In the IoT network, various type of information are exchanged. Some of them are sensitive or confidential and should be protected. IoT users should have a restricted access to this data. Most of the IoT devices are vulnerable, with limited security capabilities. They could be relatively easily controlled by malicious hackers for various attacks to send wrong data to clouds or to get access to stored information. Securing the IoT system is a paramount task that should be provided on the network side and also in the data storage system.

8.3.1.3 Storage and manipulating the huge amount of data

IoT devices have no storage capacity and low-computing resources. All the collected data is sent to remote clouds to be processed. However, the existing centralised cloud computing model is facing the scalability challenge because of the variable massive number of devices generating large amount of data. Furthermore, the distance between IoT devices and clouds, considered relatively long, may affect the QoS in the network. Because of the explosion in the volume of IoT device's data, high-efficient and sufficient computing power is required for data processing and storage and substantial applications execution. Currently, most IoT solutions are based on the centralised cloud paradigm that may work properly. However, due to the rapid growth and diversity in number of IoT devices, centralised traditional network architecture is facing new challenges to meet new and future service requirements:

- *Availability, Efficiency, and Resilience*: Centralised data recording in such huge scalable systems is tedious and introducing a single point of failure. Affected by the changeable demand for simultaneous computing, processing, and storage resources, exceeding the capability of centralised clouds, it is difficult to collect the volumes of traffic generated by a huge number of devices in a central location for timely processing and analysis while respecting the quality of experience (QoE) that should be provided by each IoT application.
- *Latency*: Because of the slow and complex procedure of simultaneously recording and processing the amount of data gathered from the huge number of IoT devices, the reaction of the cloud may cause enormous delays, affecting the performance of many applications such as applications with real-time data delivery such as video surveillance. However, this is becoming more and more difficult on a large scale that may cause a huge delay because of the limited cloud capacity with the remarkable increase in the amount of communications.

- *Security*: One of the important objectives of designing new distributed architectures is securing IoT networks. With the traditional cloud system, some pieces are missing in relation to privacy and security.

8.3.1.4 Energy consumption

Energy consumption is one of the challenges facing the IoT network. Indeed, the tiny IoT devices are equipped with batteries with limited capacity. The big number of sensors embedded in IoT devices produces a lot data that should be processed to extract meaningful information through big data analytics. These data have to be gathered over a long period to reach constructive and insightful conclusions. Communication is essential in such systems; however, it is power-consuming. Therefore, it is important to develop robust energy-aware protocols used to enhance the system performance mainly in terms of data transmission and dissemination to guarantee low-energy consumption because energy scarcity restrains network lifetime and robustness.

8.3.2 Blockchain general challenges

The implementation of blockchains poses a large number of challenges despite the fact that the key idea of blockchain is simple. This section highlights the main ones.

8.3.2.1 Storage capacity and scalability

Storage capacity and scalability are considered as a big challenge in blockchain. The list of the chain in the blockchain is always growing, at a rate of 1 MB per block every 10 minutes in Bitcoin. The storage requirements become significant especially because the full chain should store nodes that validate the transactions and blocks, called 'full nodes'. Therefore, the nodes require more and more resources, and an oversized chain effects negatively the performance. The nodes in the blockchain are expected to validate each transaction of each block. The number of transactions in a block and the time between blocks modulate the computational power required, and this has a direct effect on transaction confirmation times. Hence, the consensus protocol has a direct effect on the scalability of blockchain networks. In addition to that, other problems can occur because of the growing number of transactions such as the increasing of the network transaction fees (up to $60 per transaction in January 2018) and the increasing of the confirmation time for a transaction To conclude, the blockchain systems tend to become slower, expensive, and unsustainable for a use case. That's why scaling is becoming an inhibiting factor in a wider acceptance of blockchain systems.

8.3.2.2 Security: Weaknesses and threats

Blockchain attracts attentions for its highly anti-tampering property in decentralised networks. It does not require peers to trust each other. Yet, blockchain still faces many types of security threats which are as follow:

- *Attacks on consensus protocols*: In open network like a public block-chain, an attack can occur if a blockchain participant (attacker) is able to control more than 51% of the mining power. In this situation, the attacker can control the consensus in the network and can easily merge the created blocks into a long-term blockchain and threaten the blockchain system. However, from a technical perspective [13], it is not easy for an attacker to obtain more than 50% of the computing power.
- *Double spending*: The adversaries, having conflicting transactions, attempt to mislead the transaction receivers, and this may happen when a transaction uses the same input as another one that has already been broadcast on the network.
- *Eclipse attacks*: Eclipse attacks means that an adversary can eclipse a node, allowing it to communicate with malicious nodes and prevent it from connecting to any honest peers in a P2P network. So it can manipulate the P2P network and gain control over a node's access.
- *Vulnerability of smart contracts*: The openness and irreversibility of blockchain makes smart contracts susceptible. Bugs and frauds are transparent to the public including adversaries. An outstanding example is the attack to the decentralized autonomous organization (DAO) in 2016, known as the DAO attack, which resulted in a forked Ethereum blockchain.
- *Distributed denial-of-service attack (DDoS)*: The adversaries exhaust the blockchain resources (such as exhausting the whole network processing capability) by launching a collaborative attack.
- *Leakage of private key*: The attackers can steal the private key of an account to take over the account. This can be achieved via traditional network attacks or capturing physical nodes.

8.3.2.3 Anonymity and data privacy

Blockchain is known by the transparency of actions established within this network. Indeed, each transaction can be controlled, audited, and retraced from the system's very first transaction. This transparency contributes without doubt to the establishment of trust. However, this transparency has causal sequence on privacy, even if there is no direct relationship between wallets and individuals, user anonymity seems to be compromised despite the mechanisms that Bitcoin provides, such as the possibility of users to create any number of anonymous Bitcoin addresses that will be used in their

Bitcoin transactions as well as the use of multiple wallets. To remedy this problem, many efforts have been made to strengthen anonymity in Bitcoin. On the other hand, many applications based on public blockchain technology require a high level of privacy in the chain, especially those dealing with sensitive data.

8.3.2.4 Wasted resources

Validation and cooperation within the network often requires spending resources, such as computational power or coins. Minimizing resources or energy spent forms significant criteria for evaluating blockchain performance. Proof-of-work (PoW) consensus algorithms, for example, are known to be energy intensive as they spent significant amounts of energy to validate transactions. In an empirical analysis, [13] found that about 10% of announced new blocks on the Ethereum network were uncles (forks of length 1). This can be seen as wasteful but is just a small indication of the vast duplication of effort in PoW mechanisms. As a result, blockchain developers are increasingly moving towards proof-of-stake (PoS) schemes as alternative to the PoW. But some authors argue that in PoS, the miners also need to work out the right hash to create new blocks, and they have to wait for a certain number of blocks to confirm the transactions. In addition to that, the validation and verification of data comes with high hardware and energy costs. Adding to the cost of information verification, blockchain systems also face an additional cost of storing the data in continuously expanding ledgers. While this is a significant concern and waste of resources needs to be minimised, it is also crucial for not compromising blockchain system security. In fact, the design of validating mechanisms and incentives can determine system vulnerabilities to malicious behaviour, potential cyberattacks, or collusion. This results in a trade-off between security and waste of resources and cost.

8.3.2.5 Technological challenges: Throughput, latency, size, and bandwidth

The throughput in the Ethereum blockchain currently supports roughly 15 transactions per second, compared to the 45,000 processed by Visa. This limitation of Ethereum and other blockchain systems has long been the subject of discussion by developers and academics. Latency is also an issue. It is the time needed from the creation of a transaction until the initial confirmation of it being accepted by the network. Transaction inclusion in the absence of network congestion takes a certain amount of time. To ensure that the transaction does not get removed due to accidental or malicious forking, a number of confirmation blocks are highly recommended. That means that transactions can be seen as committed after 60 minutes on average in Bitcoin, or 3–10 minutes in Ethereum. Having low latency is crucial when designing a real-life payments system. Size and bandwidth limitations are considered as variations of the throughput issue. If the transaction volume of VISA were to be processed by Bitcoin, the full replication of the entire blockchain data structure would pose massive problems and a challenge in data storage and bandwidth. Private and consortium chains aim to address these challenges. In this context it is worth noting that most everyday users can use wallets instead, which require only small amounts of storage.

8.3.3 Fog and edge challenges

Fog and edge are merging architecture to replace the traditional cloud system. Fog nodes are distributed computing resources, located at the edge of the network, that allow the deployment of fog services. Based on different technologies, all the fog-layer devices are connected and abstract to be considered as a single logical entity defined as the 'fog node', able to perform distributed tasks, as it is on a single device. This architecture solves some issues of the traditional centralised cloud. Indeed, the existence of many fog nodes located in different areas of the network and close to the end-user devices services at the edge by providing speedy high-quality localised services with low latency and meeting the demand of real-time interactions and latency-sensitive applications. Also, it allows the reduction in traffic in the network by reducing the number of retransmissions in one direction (to the same cloud server). Moreover, all fog nodes are connected and collaborate together, and it should keep providing service normally, if some Fog nodes fail. Indeed, computation should turn quickly and continue on other nodes to provide better service availability. Therefore, this architecture is more suitable for scalable network. However, few challenges should be taken into consideration when designing fog and edge systems [14–17].

8.3.3.1 Scalability and adaptability

This architecture is used in a dynamic, scalable network. Thus, it should be able to adapt to the changing environment and to meet the variable needs and demands of scalable IoT applications like IoT. Achieving a linear performance is the challenge in such systems.

8.3.3.2 Heterogeneity, efficiency, and performance

The efficiency of this architecture is one of the most important challenges. Indeed, heterogeneous computing nodes may be used to create the fog, and this should not affect the performance of the applications. Communication between fog nodes and end-user devices (fog-fog and fog-device) should be facilitated, anytime and anywhere because many databases will be manipulated by the various fog nodes.

8.3.3.3 Security and privacy

Securing collected data is one important challenge of the fog system because most of the collected data is private, sensitive, and sometimes confidential. The data storage system should be aware of this need and provide solutions to solve this problem. Fog system faces two big issues:

- How to trust data managing entities and provide data integrity?
- How to avoid malicious attacks?

To answer these questions, we need to focus on the issues that may face the fog system:

> *Authentication and Data Integrity*: Fog nodes communicate with several entities within multiple layers (fog-to-fog, fog-to-cloud, and fog-to-end-user devices). This multi-layer collaboration system has several issues in term of security and privacy, especially with data access control, device identity management, authentication and authorisation, and securely sharing of information between the various nodes. Authentication is the most challenging security issue in the distributed fog systems because the nodes communicate and provide service for a considerable number of devices, unlike the cloud system which is based on unique central authentication server. Furthermore, the data integrity is another important requirement that should be guaranteed because the data can be processed by several fog and cloud devices. It should be analysed and distributed without reviling it. Fog architecture should support secure collaboration and interoperability between heterogeneous entities.
>
> *Malicious Attacks*: The deployment of fog nodes close to the end-user devices, where protection and surveillance are relatively weak, makes this system vulnerable to several malicious attacks, which do not exist in cloud computing such as denial of services and man-in-the middle.
>
> - *Denial of services (DOS)*: It is one of the most challenging security attacks in online services and websites. In this type of attack, the attacker blocks access to the user's target fog nodes by requesting for infinite processing and storage services. This can occur at different levels. The attacker may spoof the IP address of the end-user devices and sends multiple fake requests to the fog node requesting to access the processing and storage resources. Also, at the end-user level, the attacker may jam the wireless channel and disable communication with the fog infrastructure. This type of attack is more and more intense when several attackers simultaneously launch the attack. They may interrupt all services provided locally by the fog system.

- *Man-in-the-middle*: In this attack, an intruder can interrupt or sniff communication between fog nodes, and they can replace the fog devices.

Efficient and secure deployment of fog infrastructure to facilitate communications between these nodes and devices in the network is still an open issue because existing strategies to secure and prevent attacks are not suitable for fog computing because of the distributed and openness characteristics of this infrastructure. Operating databases in a distributed manner resolves some issues such as delay but introduces others related to authenticating and trusting the data managing entities because many fog nodes should communicate together and with end-user devices to store and manipulate data in the various databases. One of the most interesting solutions for coping with such requirements is blockchain.

8.4 BLOCKCHAIN FOR EDGE AND FOG

The IoT applications are more and more widely used. It is expected to connect an enormous number of heterogeneous devices in a smart way to provide diverse services. To connect this large number of devices and provide the requested services, IoT systems are facing many challenges in term of infrastructure and design that include network coverage, massive traffic load, security and privacy, high system availability and reliability, QoS, data storage and manipulation, and energy constraints. Many solutions are provided to solve each issue. Thus, the best way to overcome these challenges and provide highest system efficiency, modern technologies and paradigms should be included to IoT systems.

Distributed data storage and management is one promising solution to solve data storage and QoS issues such as edge and fog computing mechanisms. Moreover, blockchain is an important paradigm that can be used with distributed systems to enhance security and privacy in IoT systems. The combination of these paradigms allows IoT networks to overcome its challenges and achieve high system performance.

While thinking about integrating blockchain with IoT systems, an important question arises about the location where these interactions will take place. Fog computing can play a key role to facilitate this integration. Indeed, its distributed computation resources are in a potential place where mining of IoT initiatives can take place. Combination of these three technologies leverages the benefits of each one [6,12,18,19] (Figure 8.5).

The advantages of this alliance and their challenges are described in the two next subsections.

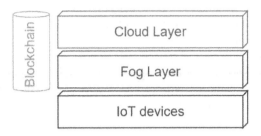

Figure 8.5 Blockchain for edge and Internet of Things (IoT).

8.4.1 Advantages

To overcome IoT challenges, new paradigms should be deployed for IoT networks. The distributed edge, fog computing, and blockchain are recent paradigms that can push IoT systems to achieve high performance and meet the main requirements of these applications. The combination of these paradigms guarantees the best system performance.

IoT-based edge and fog computing. Important technological advances have led to an expansion of IoT applications that will result in an explosion in the volume of data produced by smart devices. Thus, due to the rapid growth and diversity in the number of devices connected to the IoT, centralised traditional network architecture is facing big challenges to meet new and future service requirements. Fog and edge computing have been used to cope with these challenges to achieve better quality of service and experience by bringing the data storage and processing close to the source of data in a distributed infrastructure.

Fog computing consists of many interconnected computing sources located at the edge of the networks. Based on heterogeneous technologies, all the fog nodes are connected, aggregated and considered as a single logical entity, able to perform distributed services simultaneously. They form an intermediate layer able to communicate with the cloud when needed.

IoT-based fog computing system has several advantages:

- Important computing power able to store and process enormous volumes of data and to run substantial applications. This architecture is more suitable for scalable network such as IoT.
- Mitigation of the data traffic and overhead at the core of the network by reducing the number of retransmissions in one direction (to the same cloud server).
- Achieving higher latency efficiency: the existence of many fog nodes located in different areas of the network and close to IoT devices-rich services at the edge by providing speedy high-quality localised

services with low latency and meeting the demand of real-time interactions, and latency-sensitive applications.
- System lightness
- Better network flexibility and availability because all fog nodes are
  connected and collaborate together, it should keep providing service
  normally; if some fog nodes fail, computation should turn quickly
  and continue on other nodes.

IoT-based blockchain and edge and fog computing. Another new paradigm that can be an added value to the IoT systems is blockchain.
It can complement the IoT with secure and more reliable data system.
Blockchain technology is defined as the P2P distributed ledger used
to record approved events and transactions. In other words, it can be
described as distributed database containing all approved and shared data
among all validated users. Those users, should in turn approve the new
added information. No data should be added to the blockchain database
or interrupted without verification. Nowadays, blockchain is supporting
many applications besides the cryptocurrency systems, including communication networks.
    Deploying blockchain paradigm for IoT networks allows to manage the
decentralisation based on the distributed edge and fog units in a trustful
way by working to avoid the heterogeneous cybersecurity attacks. It helps
to achieve more scalability, to provide privacy, and to solve reliability
problems related to IoT patterns. The combination blockchain with IoT-
based edge and fog computing may achieve diverse benefits that include the
following:

- Better management of decentralised computing resource and system flexibility thanks to the distributed feature of blockchain, thus
  achieving the required level of scalability and increasing the system
  availability; and
- Improving system security by providing privacy, trust, and reliability
  and working against heterogeneous attacks that can affect IoT data
  systems.

To increase the level of these benefits and achieve high system performance,
IoT networks employ, in addition to edge and fog computing paradigms
and blockchain, software-defined networking (SDN) mechanisms. SDN
consists in controllers used for resource provisioning and orchestration
technics based on separation the forwarding plane and the control plane
to provide a dynamic network structure. The network part that is responsible for forwarding traffic is defined as the 'data plane', while the control
plane is the part that makes the decision of the traffic. The incorporation of
SDN to IoT-based blockchain and fog computing allows achieving higher

system performance in terms of network security, flexibility, management, and scalability.

8.4.2 Issues

The combination IoT with blockchain and fog computing solves many issues of the IoT system design. However, many challenges are still open and require more attention.

8.4.2.1 Security and data integrity issues

One of the biggest constraints on the design of IoT systems, that requires high demands, are cybersecurity attacks and threats. Any IoT network should provide tools and be able to work against these malicious attacks. The combination blockchain with IoT-based fog computing can provide higher security of the IoT networks in addition to many other benefits to the overall networks such as scalability and flexibility. However, this integration still may face a challenge in the reliability and integrity of data generated by IoT devices. Blockchain can check the identity of the generator of this data, can ensure that it is immutable, and also can identify the changes when data that is already corrupted arrives in the blockchain; nothing can happen, and it stays corrupted. Corrupted data can be generated in various situations, not only resulting from malicious attacks. Indeed, IoT devices may be affected by many factors such as the environment and the device's failure. These devices should be thoroughly tested before their integration to the system in the right location to avoid physical damage. IoT devices can be hacked, and many threats can affect them such as DOS and pushing the device to generate corrupted data.

8.4.2.2 Scalability and storage capacity issues

An important issue of blockchain is the storage capacity and scalability. In the context of IoT, where devices may generate gigabytes (GBs) of data in real time, this challenge is much greater although the use of edge and fog storing resources. Indeed, some current blockchain can only process few transactions at a time and are not designed to store high volume of data, causing a great delay and an important barrier and challenge for IoT. Other techniques of data compression and data lightening should be used to simplify processing and storage tasks of the huge amount of IoT data.

8.4.2.3 Processing power and time

In blockchain, each new transaction should be verified and encrypted based on several algorithms before joining the chain as a new block. IoT and fog devices are diverse with different computing capabilities, including tinny sensors not able to run any encryption algorithm. These heterogeneous devices will not be able to run the same algorithms at the same desired speed. This should be taken into consideration to provide better performance of the overall IoT system.

8.4.2.4 Legal and compliance

Combining blockchain, IoT, and fog computing is a new aspect without any legal or compliance to follow, which causes a serious issue for IoT manufacturers and service providers. It may be a big constraint to involve blockchain technology with IoT.

8.5 SECURITY IN BLOCKCHAIN-BASED FOG COMPUTING SYSTEMS: SOLUTIONS

This section presents some solution and approaches to secure IoT-based fog and edge computing using blockchain technology in various domains.

8.5.1 Smart home solution

The use of IoT, blockchain, and Fog computing has grown significantly fast in recent years. The smart home is one of the most popular applications which can improve the resident's quality of life. However, the security and privacy issues of homeowner brings emerging challenges.

In [8], the authors proposed a hierarchical architecture that consists of smart homes, an overlay network and cloud storages coordinating data transactions with blockchain to provide privacy and security. The smart home is comprised of smart devices, local blockchain, and local storage. The smart devices are all the devices located in the home. Local blockchain is a secure and private blockchain owned by one or more devices. The blockchain is decentralised, yet the local blockchain is centrally managed by its owner. The owner is responsible for adding a new device or removing an existing one by deleting its ledger. The owner can control all transactions happening at home thanks to the policy header of the local blockchain. The communication between the devices can occur only if the owner permits them by giving them a shared key based on the generalised Diffie-Hellman algorithm. The third compound of the smart home is the local storage. It is considered an optional local storage for storing data locally, and it could be a local backup drive. In addition to these three compounds, each home's miner possesses a list of public keys (PKs) to give other permissions to access the smart home data. The constituent nodes in the network could be smart home miners. A particular user may have more than one node in the overlay network. To decrease network overhead and delay, nodes in the overlay network are grouped in clusters and each cluster elects a cluster head (CH). Each CH has three attributes: PK of requesters, PK of requestees, and forward list. In some cases, a third-party service provider (SP) can access the stored data and provide certain smart services when devices in smart home store their data in cloud storage. The use of technical mechanisms such as block number, hash of stored data, and encryption enhance the security and the user authentication. A use case is discussed to explain how transactions are handled. It explains how a user has created an account in a cloud storage facility and set up permissions for her thermostat to upload data to this facility. The proposed approach was evaluated, the authors qualitatively analyse the overhead and performance of the architecture under common security and privacy threats such as threat to accessibility (DoS, modification attack, dropping attack and mining attack), threat to anonymity, and threats to authentication and access control. Qualitative evaluation highlights its effectiveness in providing security and privacy for IoT applications.

In [20], the authors proposed an architecture to counter the security threats of the smart home. Their architecture contains three tiers: smart contracts which are building in each IoT device in the smart home, the private blockchain, which is the blockchain in each smart home, and the public blockchain, which is the P2P blockchain network connected houses. The proposed architecture solves mainly the low storage and computation of IoT devices issues. The smart home tiers can secure the use's privacy. The proposed architecture designs how devices trigger actions under certain conditions. Each IoT node is considered a node in the private blockchain and stores the locally distributed ledger. To process the transaction in the private or public blockchain, each smart home deploys a local miner. The latter can store the IoT devices data and add a new device to the private blockchain. The proposed architecture decreases the time spending on the transaction. The blockchain ensures the network security. The proposed architecture ensures the confidentiality and the integrity. The decentralised network provides the confidentiality. For the integrity, the blockchain and smart contract do not change data during the transaction.

8.5.2 Smart health care

Another sector where IoT applications can be efficient is the health care. It is possible to keep track of all the medical history of a patient or also to monitor health conditions in real time. Security and privacy and computing the amount of collected data are the key issues of such a system. Thus, use of blockchain and fog computing with IoT system has attracted many researchers in the health care sector.

In [21], authors proposed a preliminary version of an IoT continuous glucose monitors (GCM)-based system. This system allows to patients with diabetes, especially those dependent on insulin (such as children and elders), to monitor blood sugar levels by collecting levels frequently and to warn them in case of dangerous situation. This helps for diagnosis, patient monitoring, or even public health actions by pushing to advanced control of the disease. This IoT application is based on fog computing system using distributed mobile smartphones that collect the data from the CGM (a small device with a sensor that takes blood sugar readings 24 hours a day) and blockchain technology to provide transparent and trustworthy manipulation of the collected data (i.e., validation, storage, and access).

CGM provides rapid warnings and takes autonomous decisions that need to be performed as fast as possible. Its sensors collect blood sugar frequently and send them to the nearest fog nodes (the patients' mobile smartphones). The fog service processes the data and decides whether to take it based on the sensed value. The taken data is sent then to the blockchain (only a preselected group of nodes can run consensus algorithm) and the remote cloud server where it is stored. Only registered and authorised users can access this data, such as doctors, nurses, or caretakers of children or the elderly patients.

Blockchain increases the transaction privacy and accelerates the transaction validation while protecting the anonymity and privacy of the patients. Moreover, the blockchain can run smart contracts, so, for example, by performing an automated purchase when it is detected that a user is going to run out of sensors. The use of fog computing allows the transfer of the cloud's computational and communication capabilities close to the sensor nodes to minimise latency, to distribute computational and storage resources, and to enhance mobility and location awareness. In [22], another IoT-based fog computing and blockchain platform was proposed to manage patients' medical records: Fog computing was used to ensure availability and performance by storing a subset of the patients' information closer to the applications and data sources, and blockchain-based strategies were used to provide the privacy required for the medical domain.

The architecture of this proposed system, presented in Figure 8.6, is composed of four layers:

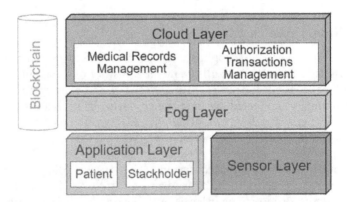

Figure 8.6 The proposed system architecture.

- *Sensor layer*: Sensing devices used to monitor the patients.
- *Application layer*: Used for manipulation and accessing patients' record. This data can be viewed by the patient (through the patient module) or by another user such as doctors or nurses through the stakeholder module, which is responsible for requesting access to view or to manipulate a subset of the patient's medical record data.
- *Fog layer*: Composed of a set of Fog nodes that functions as a blockchain miner to validate transactions such as registering patients, receiving and validating the data generated in stakeholders and sensed layer, requesting data access, granting data access, and visualizing and manipulating data. Each patient should provide the authorisation for a list of users (doctors or nurses) to view and modify specific information. Finally, this layer synchronises the subset of data and authorisations with the cloud layer.
- *Cloud layer*: Responsible for storing the complete set of patients' records and the records of authorisation to access them. It functions as the blockchain database that stores proprietary application registration transactions and authorisation transactions.

The blockchain in the proposed architecture allows fog nodes to carry out the authorisation process in a distributed way while allowing the patients to possess their own records and to control applications and users that can access and manipulate their data. Fog computing positively favours the performance of this platform, because it improves the access time to the patient's medical records by storing the information closer to the applications and devices and eliminating the single point of failure of the traditional authentication model with the cloud computing paradigm.

8.5.3 Smart vehicular

Smartness of vehicular systems is under intense development and is a timely hot topic of research. Smart vehicular systems have evolved from vehicular ad hoc networks (VANETs), the twin brother of mobile ad hoc networks (MANETs), integrating with the Internet of vehicles (IoV) [23,24] for more versatility and integration of the vehicular communication for the sake of providing the user with advanced services of assisted driving and tailored infotainment in a Smart city context. As mentioned in [25], a Smart city is conceived as an environment which is fed and monitored with advanced technologies and new types of communication paradigms bringing novel services affecting, supposedly, positively the quality of life of citizens. Authors of [25] also emphasised that multimedia communication and notably through video streaming are poised to be of high impact and a feeding factor for traffic flow as well as for providing entertainment and advertising services.

Another aspect of smart vehicular progress was illustrated in a number of works that put considerable focus on providing cloud-assisted service provisioning such as the automatic navigation as in [26] where the authors propose the design of mobile cloud computing services where vehicles or mobile devices therein play the role of both cloud server and cloud client. They describe first the architecture of vehicular networks for vehicular cyber-physical systems (VCPS) and the delay modelling for the interactive navigation and accident prediction. Then, they propose two smart road services provided through the cloud, such as interactive navigation service for the efficient driving and pedestrian protection service against road accidents.

In [27], the authors confirm the evolution of the IoV from existing VANETs, thanks to the advent of the IoT. IoV is allowing the provision of diverse types of services to drivers by means of integration of vehicles, sensors, and mobile devices into a global network. It was also emphasised in [27] that vehicular cloud (VC) will be an important medium of data collection and dissemination that is vital for intelligent transportation systems (ITS) to serve online travel systems. They have shown in their work that with the contribution of a small segment of the fleet of vehicles in a dynamic VC was enough to feed the ITS with useful and meaningful data collection.

The researchers of [28–30] also suggested the relevance of the integration of IoT-based vehicular technologies in Smart city applications for enriching service provisioning that will eventually ignite the proliferation of exciting and even more advanced technological marvels. Hence, they have proposed a framework for architectural and communication design that will serve to integrate vehicular networking clouds with IoT, which they refer to as VCoT. This framework is intended to facilitate the development of new applications catering for various IoT services through vehicular clouds. In their proposal, the authors emphasise the use of the long-range wide area network (LoRaWAN)-based [31] vehicular networking as the underlying communication architecture. Another contribution took the solution closer to the vehicular domain by proposing fog-based assistance for service provisioning. Indeed, the authors of [32] have proposed a collaborative, fog-storage-based routing protocol for video streaming over vehicular ad hoc with the aim of minimizing instances of content sharing. At the heart of the routing function is an indicator, generated by combining the speed, location, and recording angle parameters of each vehicle involved in vehicular collaboration that helps reduce the unnecessary exchange of video data in vehicle-to-vehicle communications.

Moreover, again confirming the continuous evolution of smart vehicular to embrace the convergence of mobile Internet and traditional IoT into the broader realm of the IoV. We are witnessing, currently, the IoV encompassing a sizable multi-dimensional network of interactions that involves many players with various aspects of collaboration and cooperation between entities. The IoV technology refers to dynamic mobile communication systems or models that communicate between vehicles and other objects using vehicle-to-vehicle (V2V), vehicle-to-road (V2R), vehicle-to-infrastructure (V2I), vehicle -to-building (V2B), vehicle-to-home (V2H), vehicle-to-everything (V2X), and vehicle-to-grid (V2G) interactions [23]. In this regard, the authors of [23], after surveying the state of the art in current and emerging IoV paradigms and communication models with emphasis on Smart cities deployment domains, proposed a new universal architecture for modelling IoV deployment in Smart cities that is based on a seven-layer model. The model includes a vehicle identification layer, object layer, inter-intra devices layer, communication layer, servers and cloud services layer, big data and multimedia computation layer, and application layer. Moreover, blockchain has been used to securing connected and autonomous vehicles communications [33].

As a recap of this section, we confirm that smartness of the vehicular system is coming from several aspects. It is coming from the merging of an amalgam of technologies, where the VANET was the ultimate start. Then, VANET evolved into including some infrastructure that comprised roadside units and BTS towers, along with the V2V direct communication. Then, the advent of cloud extended the capabilities of the underlying communication architecture with computing backups. Then the evolution was geared towards providing closer, fog-storage-based backup for again storage and computing platforms. The advent of IoT has ignited the desire for specialisation (as it is the fashion) to lead to the creation of the IoV that has been defined in another metamorphosed fashion into a VCoT. Throughout this myriad of technologies, architectures, and platforms, service provisioning through smart apps is the ultimate ripe fruit that will make the harvest well rewarding for developers in the near future. However, as mentioned, the true success of these apps over these platforms will not be genuine unless the security puzzle is solved. The blockchain technology as presented seems to be the winning horse on which to bet for solving this puzzle to a certain extent.

8.6 CONCLUSION

IoT has revolutionised the world by introducing small applications with roles to facilitate and enhance services and well-being. This chapter has shown the burden from relying only on cloud storage, the technologies arise to ease this burden, and their emerging issues and challenges. It also discusses that such advanced technologies require advanced solutions to deliver the quality of service and retain clients' satisfaction. Therefore, it presents blockchain technology as a viable solution for achieving secure storage in fog computing. IoT-based fog system may face crucial challenges in terms of security, privacy, and integrity of the collected data.

Indeed, in IoT, diverse types of data are exchanged, some of which are sensitive and confidential and require a high level of security and privacy. Blockchain is a P2P distributed ledger used to record approved events and transactions among all validated users. Blockchain provides an important level of security for the added data because no one can modify it and offers anonymity and privacy for the users. This combination provides an excellent solution for IoT fog applications with higher performance in several sectors: smart homes, health care, and smart vehicles, to name a few.

REFERENCES

1. Ismaeel Al Ridhawi, Moayad Aloqaily, and Azzedine Boukerche. Comparing fog solutions for energy efficiency in wireless networks: Challenges and opportunities. *IEEE Wireless Communications*, 26(6):80–86, 2019.
2. Mohammed Al-khafajiy, Thar Baker, Hilal Al-Libawy, Zakaria Maamar, Moayad Aloqaily, and Yaser Jararweh. Improving fog computing performance via fog-2-fog collaboration. *Future Generation Computer Systems*, 100:266–280, 2019.
3. Hesham El-Sayed, Sharmi Sankar, Mukesh Prasad, Deepak Puthal, Akshansh Gupta, Mamoranjan Mohanty, and Chin-Teng Lin. Edge of things: The big picture on the integration of edge, IoT and the cloud in a distributed computing environment. *IEEE Access*, 6:1706–1717, 2018.
4. Venkatraman Balasubramanian, Faisal Zaman, Moayad Aloqaily, Saed Alrabaee, Maria Gorlatova, and Martin Reisslein. Reinforcing the edge: Autonomous energy management for mobile device clouds. In *IEEE INFOCOM 2019-IEEE Conference on Computer Communications Workshops (INFOCOM WKSHPS)*, pp. 44–49. IEEE, 2019.
5. Pengfei Hu, Sahraoui Dhelim, Huansheng Ning, and Tie Qiu. Survey on fog computing: Architecture, key technologies, applications and open issues. *Journal of Network and Computer Applications*, 98, 2017.
6. Tiago M. Fernández-Caramés and Paula Fraga-Lamas. A review on the use of blockchain for the Internet of Things. *IEEE Access*, 6, 32979–33001, 2018.
7. Lewis Tseng, Liwen Wong, Safa Otoum, Moayad Aloqaily, and Jalel BenOthman. Blockchain for managing heterogeneous Internet of Things: A perspective architecture. *IEEE Network*, 34(1):16–23, 2020.
8. Ali Dorri, Salil Kanhere, Raja Jurdak, and Praveen Gauravaram. Blockchain for IoT security and privacy: The case study of a smart home. *IEEE International Conference on Pervasive Computing and Communications Workshops (PerCom Workshops)*, Kona, HI, pp. 618–623, 2017.
9. Saniya Zafar, Sobia Jangsher, Ouns Bouachir, Moayad Aloqaily, and Jalel Ben Othman. QoS enhancement with deep learning-based interference prediction in mobile IoT. *Computer Communications*, 148:86–97, 2019.
10. Faizan Ali, Moayad Aloqaily, Omar Alfandi, and Oznur Ozkasap. Cyberphysical blockchain-enabled peer-to-peer energy trading. *IEEE Computer*, 2020. https://arxiv.org/abs/2001.00746.

11. Marc Pilkington. *Blockchain Technology: Principles and Applications*, Research Handbook on Digital Transformations, edited by F. Xavier Olleros and Majlinda Zhegu. Edward Elgar, 2016. Available at SSRN: https://ssrn.com/abstract=2662660.
12. Ammar Muthanna, Abdelhamied Ateya, Abdukodir Khakimov, Irina Gudkova, Abdelrahman Abuarqoub, Konstantin Samouylov, and Andrey Koucheryavy. Secure and reliable IoT networks using fog computing with software-defined networking and blockchain. *Journal of Sensor and Actuator Networks*, 8(1):15, 2019.
13. Moayad Aloqaily, Azzedine Boukerche, Ouns Bouachir, Fariea Khalid, and Sobia Jangsher. An Energy Trade Framework Using Smart Contracts: Overview and Challenges. *IEEE Network 2020*. doi:10.1109/MNET.011.1900573.
14. Mithun Mukherjee, Rakesh Matam, Lei Shu, Leandros Maglaras, Mohamed Amine Ferrag, Nikumani Choudhury, and Vikas Kumar. Security and privacy in fog computing: Challenges. *IEEE Access*, 5:19293–19304, 2017.
15. Noshina Tariq, Muhammad Asim, Feras Al-Obeidat, Muhammad Farooqi, Thar Baker, Mohammad Hammoudeh, and Ibrahim Ghafir. The security of big data in fog-enabled IoT applications including blockchain: A survey. *Sensors*, 19:1788, 2019.
16. Ouns Bouachir, Moayad Aloqily, Lewis Tesng and Azzedine Boukerche. Blockchain and Fog Computing for Cyber-Physical Systems: Case of Smart Industry. *IEEE Computer*, 2020. arXiv:2005.12834.
17. Y. Liu, J. E. Fieldsend, and G. Min. A framework of fog computing: Architecture, challenges, and optimization. *IEEE Access*, 5:25445–25454, 2017.
18. Ana Reyna, Cristian Martín, Jaime Chen, Enrique Soler, and Manuel Díaz. On blockchain and its integration with IoT. Challenges and opportunities. *Future Generation Computer Systems*, 88:173–190, 2018.
19. Imran Makhdoom, Mehran Abolhasan, Haider Abbas, and Wei Ni. Blockchain's adoption in IoT: The challenges, and a way forward. *Journal of Network and Computer Applications*, 125:251–279, 2019.
20. Yiyun Zhou, Meng Han, Liyuan Liu, Yan Wang, Yi Liang, and Ling Tian. Improving IoT services in smart-home using blockchain smart contract. *2018 IEEE International Conference on Internet of Things (iThings) and IEEE Green Computing and Communications (GreenCom) and IEEE Cyber, Physical and Social Computing (CPSCom) and IEEE Smart Data (SmartData)*, Halifax, NS, Canada, pp. 81–87, doi:10.1109/Cybermatics_2018.2018.00047.
21. Tiago Fernández-Caramés, and Paula Fraga-Lamas. Design of a fog computing, blockchain and IoT-based continuous glucose monitoring system for crowdsourcing mhealth. *Proceedings, MDPI*. p. 37, 2018.
22. Cícero A. Silva, Gibeon S. Aquino Jr, Sávio R. M. Melo and Dannylo J. B. Egídio, A fog computing-based architecture for medical records management. *Wireless Communications and Mobile Computing*, 1–16, 2019. doi:10.1155/2019/1968960.
23. Gerald K. Ijemaru, Adamu Murtala Zungeru, Li-Minn Ang, and Kah Phooi Seng. Deployment of IoV for Smart cities: Applications, architecture, and challenges. *IEEE Access*, 7:6473–6492, 2019.

24. Talal Ashraf Butt, Razi Iqbal, Khaled Salah, Moayad Aloqaily, and Yasser Jararweh. Privacy management in social Internet of Vehicles: Review, challenges and blockchain based solutions. *IEEE Access*, 7:79694–79713, 2019.
25. Emna Bouzid Smida, Sonia Gaied Fantar, and Habib Youssef. Video streaming challenges over vehicular ad-hoc networks in smart cities. 2017. International Conference on Smart, Monitored and Controlled Cities (SM2C), Sfax, 2017, pp. 12–16, doi:10.1109/SM2C.2017.8071838.
26. Eunseok Jeong, and Jaehoon Lee. VCPS: Vehicular cyber-physical systems for smart road services. *2014 28th International Conference on Advanced Information Networking and Applications Workshops*, Victoria, BC, 2014, pp. 133–138, doi:10.1109/WAINA.2014.146.
27. Moumena Chaqfeh, Nader Mohamed, Imad Jawhar, and Jie Wu. Vehicular cloud data collection for intelligent transportation systems. *2016 3rd Smart Cloud Networks & Systems (SCNS)*, Dubai, 2016, pp. 1–6. doi:10.1109/SCNS.2016.7870555.
28. Hasan Ali Khattak, Haleem Farman, Bilal Jan, and Ikram Ud Din. Toward integrating vehicular clouds with IoT for smart city services. *IEEE Networks*, 33:65–71, 2019.
29. Ismaeel Al Ridhawi, Moayad Aloqaily, Burak Kantarci, Yaser Jararweh, and Hussein T. Mouftah. A continuous diversified vehicular cloud service availability framework for smart cities. *Computer Networks*, 145:207–218, 2018.
30. Moayad Aloqaily, Burak Kantarci, and Hussein T. Mouftah. Multiagent/ multiobjective interaction game system for service provisioning in vehicular cloud. *IEEE Access*, 4:3153–3168, 2016.
31. Lorawan Alliance. https://lora-alliance.org/about-lorawan, December 2019.
32. Allan Douglas, Hugo Santos, Denis Rosario, Paulo Bezerra, Adalberto Melo, and Eduardo Cerqueira. A collaborative routing protocol for video streaming with fog computing in vehicular ad hoc networks. *International Journal of Distributed Sensor Networks 2019*, 15(3):15, 2019.
33. Geetanjali Rathee, Ashutosh Sharma, Razi Iqbal, Moayad Aloqaily, Naveen Jaglan, and Rajiv Kumar. A blockchain framework for securing connected and autonomous vehicles. *Sensors*, 19(14):3165, 2019.

Chapter 9

Differential privacy for edge computing-based smart grid operating over blockchain

Muneeb Ul Hassan, Mubashir Husain Rehmani, and Jinjun Chen

CONTENTS

9.1 INTRODUCTION

Blockchain emerged as one of the most appealing technologies of this decade because of its promising features [1]. First, blockchain was proposed to serve a community in the form of a cryptocurrency named Bitcoin, but later the decentralised nature of blockchain attracted the attention of both industry and academia. The main reasons for the attention are the transparent, tamper-proof, and secure decentralised nature of it. Blockchain technology works on the concept of a decentralised distributed ledger in which transactional records are store; these records are immutable and can be verified at any time. Every node of blockchain has a copy of that distributed ledger and can personally verify the transactions to ensure that everything is fair [2]. The data is added to blockchain with the help of a consensus mechanism that requires approval of certain nodes in the network; therefore, the possibility of false data in the network is reduced to a minimum. This specific process of data addition is known as 'consensus' and is carried out with the help of specific nodes known as 'miners'. Plenty of other fields such as smart grid, healthcare, and others that require dealing with users' data are adopting blockchain technology to enhance trust in the network. Much research shows that the integration of blockchain in such fields have produced tremendous results. However, certain processes involved in blockchain implementation require high computational complexity, such as mining, carrying out consensus, etc. Therefore, it is not possible to integrate blockchain directly with these applications. To overcome this issue, edge computing emerged as one of the most promising platforms to solve the problem of computational complexity to maximum extent for such blockchain nodes.

Edge computing, also known as fog computing, is an extension of the cloud which was introduced to the enhance quality of service (QoS) for cloud applications by reducing delays in computing-expensive tasks [3]. Edge computing effectively provided computational power, application services, and data-storage capabilities to its users along with serving as a bridge between user and cloud storage. On the other hand, this edge computing technology also lacks security and is vulnerable to security threats, such as sniffer attacks, jamming attacks, etc [4]. These issues can be solved by integrating it with blockchain technology because of the cryptographic security and immutable nature of blockchain. Thus, this integration of blockchain and edge computing is beneficial for both in their own terms because it enhances both the security and computational power for edge nodes and blockchain nodes, respectively.

This concept of blockchain-based edge computing is being used in many applications such as healthcare and smart grid. One of its subapplication is a real-time smart meter reporting using an advanced metering infrastructure (AMI) network. In an AMI network, smart meter users (usually house owners) report their real-time data to the grid utility with the help of a local aggregator. Before the integration of blockchain, smart meter nodes used to report their real-time readings to local edge aggregator to carry out certain computations, such as bills, and then return the results to smart meter users. One major issue in this system was lack of trust on the local aggregator and grid utility because of the absence of transparency as aggregator or grid utility can tamper with the calculated readings or can play with the values according to their choice without letting smart meters know about it [5]. To overcome this, blockchain is one of the most viable solutions. Currently, in the proposed scenario, smart meters will act as blockchain nodes, and data will be available at each node so that the grid utility or local aggregator cannot alter anything. This solves the issue of trust at local aggregator; however, the issue of privacy is still present. As local aggregator has to share readings of smart meters to the grid utility, usually the grid utility requires on-demand readings to carry out certain computational works such as load forecasting, etc.

The local aggregator, being an edge blockchain node reports the stored values to the smart grid utility to perform such tasks, but the reported values contain the private information of smart meters users, which should not be leaked. For example, if the grid utility asks following queries from local aggregator to know some personal information 'How many houses are currently producing energy from biomass?' 'Which houses are capable to produce energy from biomass?' and 'How much energy is being produced from biomass at a specific time?' Let's suppose that only one house has biomass energy production unit, and if local aggregator answers all these queries correctly, it can leak a user's privacy. Therefore, the protection of such instantaneous on-demand reporting is required. To solve this issue, we use the phenomenon of differential privacy. Differential privacy ensures that the observer cannot get exact private information because of its perturbation mechanism [6]. Therefore, the integration of differential privacy with a blockchain-based smart grid can protect user privacy along with reporting on-demand readings to the grid utility.

In this chapter, we propose a blockchain-based differentially private on-demand edge reporting phenomenon (BDOR) for smart grid AMI networks. Our proposed strategy ensures that private data of smart meters gets protected even in case of on-demand query evaluation by the smart grid. Moreover, the privacy criteria of our proposed work can be adjusted depending on the requirement of privacy and utility. So, there will be a trade-off between utility and privacy that can be adjusted by carrying out negotiations and discussions between smart meter users and the grid utility. Performance evaluation of BDOR mechanism shows that our proposed work efficiently reports smart meter data queries without compromising the privacy of its users.

The key contributions of our work are as follows:

- We provide a comprehensive analysis of the integration of privacy in blockchain and edge computing applications.
- We provide an edge computing-based blockchain model for smart grid AMI network.
- We integrate the phenomenon of differential privacy in blockchain-based edge computing scenario.
- We propose BDOR privacy preservation mechanism that efficiently protects on-demand readings of smart meter users from the grid utility.
- We develop an algorithm for BDOR strategy for on-demand reporting and effective privacy preservation.

To the best of our knowledge, no work provides a detailed analysis over privacy preservation in blockchain and edge computing along with proposing an efficient mechanism for private on-demand reporting (Table 9.1).

9.1.1 Comparison of privacy-preserving techniques

The field of privacy preservation is not new and researchers have searched for the most optimal way to preserve privacy. In the modern technological world, it started from sending encoded messages; now we have intelligent privacy-preserving strategies that can make decisions themselves. In this subsection, we discuss four major privacy-preserving strategies (i.e., anonymisation, encryption-based privacy, swapping, and differential privacy) of the current era along with mentioning their advances, features, drawbacks, functioning, and many other properties.

1. *Anonymisation:* Let us begin the discussion with one of the pioneering privacy-preservation strategies for statistical databases, *Anonymisation* [17]. Anonymisation can be defined as a method of protecting statistical databases during analysis by removing personally identifiable information (PII) from them. This PII can vary from the privacy requirement to the type of databases (e.g., in some databases age could be PII, but in others it might be termed as PII.) So, finding exact PII that ensure maximum privacy has always remained a challenging task for researchers. To date, plenty of anonymisation techniques have been proposed by researchers ranging from k-anonymity [7] to onion routing [18]. However, k-anonymity [7], l-diversity [8], and t-closeness [9] are still the three most famous anonymisation-based privacy strategies. Anonymisation works on the principle of finding and removing direct identifiers that can further be used to invade the privacy of users. No doubt, anonymisation remained an effective technique for many years in the field of privacy

Table 9.1 Comparative analysis of privacy-protection strategies on the basis of mechanism, base idea, variant examples, drawbacks, and possible attacks

Name of privacy	Mechanism functioning	Base idea	Variant examples	Drawbacks	Possible attacks
Anonymisation	Database privacy is protected by removing personally identifiable information (e.g., date of birth, ID number, etc.) before conducting surveys	Finding direct identifiers	• k-anonymity [7] • l-diversity [8] • t-closeness [9]	• Loss of original data • Reidentification chances exists • Finding exact PII	• De-anonymisation attacks • Background knowledge based-attacks • Data correlation attacks
Swapping	Protecting dataset by exchanging values of some individuals	Finding swapping subsets	• Equi-width [10] • GroupSwap [11] • Value swapping [12]	• No formulated theoretical definition	• Data correlation • Adversarial learning attack
Encryption-based privacy	Use of public and private keys in order to protect individual privacy from third-party intruders	Key generation and distribution	• Homomorphic encryption [13] • Identity-based cryptography [14]	• Computationally expensive	• Cipher text attacks • Known message attack
Differential privacy	Adding random noise value to perturb output value in case of both; real-time and statistical data	Finding optimal epsilon and delta values	• Random-dp [15] • Rényi-dp [16]	• Privacy-utility trade-off	• Correlation attack • Background knowledge attack

Abbreviations: dp, differential privacy; PII, personally identifiable information.

preservation, but as a result of the advancement in modern machine learning algorithms, attackers have strengthened and now protecting a database by anonymising the databases is not enough. Attackers can carry out data correlation attacks and link attacks to invade privacy and can easily de-anonymise data by relating it with other anonymised data sets. Therefore, researchers are trying to integrate anonymisation with some other privacy-preservation strategies to ensure complete privacy.

2. *Swapping:* Another well-known privacy-preserving strategy that is being used by researchers to protect statistical and real-time privacy is swapping [19]. In swapping, randomly selected data values are swapped with each other to ensure that the output database becomes anonymised and no adversary can link data records with some individual with confidence. One major hurdle that is faced by researchers during this implementation is to find the most adequate swappable subsets that will provide maximum privacy. This technique is not developed much because of lack of theoretical background, but plenty of variants have been discussed in literature, for example, equi-width [10], GroupSwap [11], and ValueSwap [12].

Moving further to some drawbacks, swapping still does not have enough theoretical background, and no one can assure the level of privacy that a swapping algorithm will provide. To eradicate this issue, some researchers integrated it with differential privacy or evaluated it according to privacy rules of differential privacy. Furthermore, certain attacks such as adversarial learning attacks are still a nightmare for swapping strategies. Although, swapping based protection is increasing slowly, there is still much that needs to be covered.

9.1.2 Encryption-based privacy

Sending private messages has been in discussion for a long time, and many privacy-protecting cryptographic messages have been discussed in the literature. However, in this study we will discuss encryption-based privacy strategies such as homomorphic encryption, which allow third parties to carry out their required operations without risking the privacy of the database. This is a privacy-preserving strategy that allows carrying out of mathematical computations on encrypting data without compromising the privacy and encryption of database [13]. To simplify it further, for example, person X wants to perform a multiplication tasks on two numbers, N_1 and N_2, but the person does not know how to multiply those numbers. On the other hand, a company or individual, Y, knows how to carry out that multiplication, but person X does not trust Y and does not want to share the two numbers. So, X encrypts the number N_1 and

N_2 into two other numbers N_3 and N_4 and sends the new numbers to Y. Similarly, Y, which only has access to N_3 and N_4 performs $N_3 * N_4$ and transmits the result back to X. After getting answer, X decrypts the answer and gets the desired result. This can be termed a mechanism in which one can perform operations over the encrypted data to get encrypted results that does not reveal privacy.

Despite being so useful, encryption-based privacy preservation is not in practice much because of the computationally expensive nature of these encryption strategies. Because to perform even a basic task, one has to generate a set of private and public keys, so that afterward encryption, distributing these keys, and decrypting the messages also come up as hectic tasks. Additionally, the scope of these encryption strategies are limited (e.g., can perform mathematical operations, etc.), and they cannot widely be applied to modern domains. Furthermore, certain cryptography attacks such as ciphertext attacks and known message attacks make the practical implementation more difficult. However, encryption-based privacy is difficult to implement, but still it has large scope if researchers can solve these issues.

1. *Differential Privacy:* Differential privacy is currently one of the most valued privacy-preservation strategies because of its dynamic nature and strong theoretical guarantee. From differential privacy, one can precisely estimate the amount of privacy leakage because of its strong mathematical analysis. Differential privacy works on the phenomenon of adding random noise to protect privacy, which was first introduced by C. D work to protect statistical data privacy [20]. However, in time, researchers further applied differential privacy in real-time scenarios, and it provided fruitful results. To date, vast literature is available in the field of differential privacy and many variants have been proposed in this field. More detailed discussion about variants of differential privacy is provided in Section 9.2.5. Some drawbacks of differential privacy is that sometimes it becomes hard to manage the trade-offs between privacy and utility. Because differential privacy is a noise addition mechanism, large noise can destroy the utility and data can become useless. Therefore, researchers are actively working to find the most optimal mechanism in this regard.

The remainder of this chapter is organised as follows: In Section 9.2 preliminaries of BDOR strategy are discussed. Section 9.3 provides a detailed analysis of privacy protection in blockchain and edge computing systems. In Section 9.4, the description of BDOR strategy along with its functioning and system model is presented. Section 9.5 contains performance evaluation of the BDOR strategy. Finally, Section 9.6 presents conclusion along with some discussion about future work.

9.2 PRELIMINARIES OF PRIVATE SMART GRID EDGE COMPUTING OVER BLOCKCHAIN

Privacy can be interpreted as a method to protect unwanted sharing of data. Similarly, another definition of privacy signifies that particular individuals should have control over their data in a way that they can control. To do so, many privacy-preserving strategies have been developed by researchers, and differential privacy is one of the top-ranked privacy-preservation strategies to protect real-time data of decentralised edge computing-based smart grid from intruders. In this section, we discuss some preliminaries of our proposed strategy.

9.2.1 Smart grid

Climate change and an increase in greenhouse gases have captured the attention of researchers working in the field of energy systems. One possible solution proposed to overcome these issues is the maximum possible use of renewable energy resources (RERs) in daily-life scenarios. This is a prospective solution; however, the integration of RERs with traditional energy system is not possible and this gave rise to a completely new energy management system, smart grid [21]. Modern smart grid technology has the potential to successfully incorporate all the characteristics of the traditional grid, along with providing a state-of-the-art interface for integration, control, handling, and management of modern grid operations. Another important component of modern smart grid is the smart meter. Smart meters are capable of performing multiple operations, such as real-time reporting, billing, value-added services, domestic RERs management and handling, etc. Primarily, an automated smart meter is assigned to carry out four major tasks such as measuring energy consumption of smart homes with variable time, report this measured reading to grid data management utility for statistical operations, receive information regarding pricing and direct control command, and exchange information with appliances of smart home to optimise their use [22]. The data collected by these modern smart meters fulfils the need of information source in the smart grid; therefore, smart meters serve as one of the most critical sources of energy data. This data is further used to perform certain analysis operations to build efficient grid systems. However, this real-time reporting is not always beneficial and can serve as for the unintended purpose of breaching the privacy of smart home consumers. A vast amount of literature shows that individual appliance data can be extracted after analysing load usage data of smart homes. Furthermore, the real-time reporting also provides information about occupancy and unoccupancy of a smart home, which reveals personal lifestyle of residents. For example, authors in [23] showed that even displayed TV channels can be estimated from electricity using profiles with a sampling time of 0.5 seconds. Moreover, the threat of smart meters is not just limited to privacy; smart meters also have some security threats, such as network security attacks and physical layer security attacks. That can lead to leaking the private data of a smart home user. Therefore, protecting the privacy and security of smart meters are the major issues that need to be tackled in modern smart grid systems.

9.2.2 Fog and edge computing

For the last 10 years, an immense increase in cloud-based services has been seen and storage-rich cloud resources are being used to provide network management and storage to its users. A lot of data giants such as Amazon, eBay, and others are using cloud-based services to store and manage their customers' data [24]. This trend is not just limited to certain companies; almost every sector benefits from cloud-based services and the smart grid is one of them. Data reported from smart meters is usually stored in cloud platforms after passing via some aggregator. However, because of massive load over cloud servers, a new trend of having a middle-ware node, edge or fog is trending now. The concept of edge computers was introduced to facilitate certain delay-sensitive applications, for example, medical data extraction, virtual reality, etc. Although, this technology was so promising that almost every cloud-computing industry started shifting its functioning via fog computing. Generally, edge nodes can be considered intermediaries between core cloud node and edge node (at user level). These nodes provide more timely responses by reducing computational latency. Furthermore, these edge nodes do also support information and data flow and direct data towards its specific destination, and hence, reducing the burden from the end node device (such as mobile phones) [25]. The goal of these systems is to provide an architecture that is capable of carrying out delay-sensitive and computing-expensive tasks at the edge of the network. Similarly, in context of the modern smart grid, edge computing also provides an infrastructure to perform delay-sensitive tasks such as on-demand reading collection, etc. Usually, the aggregator serves as the edge computing node by providing privileges to smart homes to interact with only the prescribed aggregator in the time of delay-sensitive or computing-expensive tasks. Therefore, the integration of edge nodes with a cloud-computing-enabled smart grid is considered to be one of the new paradigms in the field of electric smart grid, and many of smart grid utilities are modifying their setups to adopt this new paradigm.

9.2.3 Blockchain

Blockchain emerged as a revolutionising technology after S. Nakamoto founded Bitcoin as cryptocurrency in 2008 [26]. Blockchain appeared as decentralised distributed ledger in which append-only data records are distributed among every node in a peer-to-peer (p2p) network. Generally speaking, blockchain consists of 7 layers which are the data layer, network layer, consensus layer, ledger topology layer, incentive layer, contract layer, and application layer [4]. The individual functionalities of these layers do not vary much and remain similar, although some variations can be found in accordance with the applied domain. For example, the core function of data layer is to encapsulate and store data frames generated from transactions of various blockchain nodes. Transactions from blockchain nodes are packed into blocks after verification, and these blocks are linked with each other with the help of hashes. So, the data layer is responsible for the storage and management of these blocks. Furthermore, the network layer is responsible for carrying out various networking operations, such as the propagation of messages to all peers and the successful dissemination and broadcast of blocks after any update. Once a transaction gets completed, it is the responsibility of the network layer to disseminate this transaction to every blockchain node within a given time frame.

Afterward, a consensus mechanism is carried out that ensures that only legal and legitimate blocks get mined in the network, and the layer taking responsibility of consensus mechanisms is the consensus layer. Another layer involved in all this process is the incentive layer, which integrates certain economic factors together and deals with all economic incentives within the network. Another layer recently added in the blockchain network is the contract layer, which is named after 'smart contract' phenomenon being used in modern blockchains. This layer provides programmability function in blockchain, so that the users or the governing authorities can make their decisions according to their choices and needs and can modify the functioning without much hassle. Finally, the last layer in the blockchain network is the application layer, which is comprised of IoT nodes or end devices. Blockchain users have access to this layer via their end devices such as smartphones and other home gadgets. This complete architecture forms an immutable decentralised storage known as a blockchain.

Blockchain as a network provides a secure platform to carry out multiple user-level operations, such as smart metering, etc., however, the computational power of certain IoT devices is not enough to mine a block independently. Therefore, research has suggested integrating edge computing with blockchain-based applications. This integration solved certain security and computational issues, although this network is not completely private and has a lot of privacy issues because of its transparent nature. Therefore, the integration of a privacy-preservation strategy with smart grid operating over blockchain-based edge computing scenario is important and needs to be considered before its practical implementation.

9.2.4 Differential privacy

Differential privacy was first introduced by C. Dwork as a notion to protect statistical data privacy. The mathematical formulation of differential privacy proved to be of great benefit, and researchers started applying differential privacy in various domains [27,28]. Differential privacy works because of its basic definitions that demonstrate that an adversary cannot predict the presence or absence of an individual in a data set with confidence. Another formal definition of differential privacy with respect to two neighbouring data sets is as follows:

1. *Definition 1: Neighbouring Differential Privacy:* A random algorithm of a function will satisfy the condition with respect to the occurrence probability if for two provided data sets, and, the outcome is in a range, we get:

$$O_r\left[X\left(a^1\right) \in O_c\right]\exp(\varepsilon) \le O_r\left[X\left(a^2\right) \in O_c\right] \tag{9.1}$$

   This equation is the output range of resultant function. Similarly, the symbol represents privacy parameter that is used to determine the level of privacy required for a specific mechanism.

2. *Definition 2: Global Sensitivity:* Global sensitivity in differential privacy works on the limit of maximum difference value between the results of two same queries asked from two independent data sets differing with each other by just one element. If a query Q_i is asked from two independent databases, B_1 and B_2, then the value of global sensitivity ΔF_q can be found via the following formula [29]:

$$\Delta F_q = \max_{B_1, B_2}\left\|Q_i\left(B_1\right) - Q_i\left(B_2\right)\right\| \tag{9.2}$$

   a. *Query:* A query can be termed a curious question asked by an analyst to a statistical database containing private information.

3. *Definition 3: Query:* A query, Q, is a curious question that takes a specific database (D) as input and generates the results according to the given function $Q(D)$ [30]. Queries can further be classified into numerical, vector, and non-numerical queries, depending on their output result requirement. Usually the Laplace mechanism deals with numerical queries, and exponential mechanism deals with non-numeric queries.

For example, an analyst is given query access to a smart grid dataset, D: The 'How many houses are generating renewable energy more than X kWh in suburb Y?' is a type of numerical query because it takes D

database as input and returns the total number of smart houses. It is not necessary for the output to be a numeric value or number. It can also be vector, for example, 'list of houses having solar as primary source?' or it could also be a non-numeric value such as owner names, etc.

Apart from statistical databases, differential privacy has also been applied to real-time data such as smart metering and healthcare data reporting, and it has provided fruitful results [27]. Therefore, in this work, we integrate differential privacy with smart grid operating over a blockchain-based edge network to protect the privacy of smart grid users (Table 9.2).

9.2.5 Variants of differential privacy

Differential is considered a formal definition of privacy that can be used to compute the level of disclosure a privacy mechanism will provide [39]. Since the first introduction of differential privacy in 2006, many variants of differential privacy have been introduced by researchers to enhance it. Some include providing a strict definition of privacy, while some used different types of noise addition mechanisms and guarantees to provide advanced protection. In this chapter, we classify 10 different variants of differential privacy on the basis of their functioning, privacy criterion, noise addition mechanism, etc. The detailed description of these mechanisms is provided in Table 9.2. Some researchers worked on the integration of differential privacy with other similar privacy-preservation strategies or other techniques that can further help in improvement of differential privacy. One such work has been carried out by Cheu et al. [31]. In this work, the authors considered the issue of development of robust and scalable privacy-preserving mechanisms on the criteria of differential privacy and developed a shuffling-based differential privacy mechanism. The proposed also named distributed differential privacy via shuffling-use mixnets 'shuffler', which serves as a cryptographic primitive to implement a differentially private algorithm. Afterward, the authors algorithmically compared the proposed algorithm to that of baseline differential privacy. Another work in the generalisation of differential privacy for text document processing was carried out by the authors in [32]. The authors addressed the problem of considerable text obfuscation by eradicating stylistic clues in the text files that can further be used to infer into private data. The proposed work integrated machine learning over 'bag or words' to protect privacy via differentially private criterions. The authors used the Laplace density function to obfuscate texts in the document. Furthermore, the authors replace words with other words with respect to Earth mover matrix to ensure privacy. The proposed strategy is evaluated over the fan fiction data set, and the results showed that the generalised differential privacy strategy is one of the most suitable mechanisms of protecting privacy.

Table 9.2 Comparison of variants of differential privacy on basis of functioning, noise addition mechanism, privacy criterion, key functioning, and evaluation type

Mechanism name	Ref #	Year	Major functioning	Noise addition distribution	Privacy criterion	Key functioning	Evaluation type
Distributed DP	[31]	2019	Used mixnet as shuffler which serves as a cryptographic primitive to implement dp algorithm. Afterward, algorithmically proved shuffler to be differentially private	Shuffling, instead of noise addition	(ε, δ)-satisfies	Shuffling via randomiser	Accuracy analysis
Generalised DP	[32]	2018	Generalised DP mechanism integrated with machine learning is applied over 'bag of words' to protect privacy. Replace words with other words with respect to Earth mover matrix	Laplace density function	(εD_x)-privacy	Generate private bag from bag of words	Epsilon on set
Rényi-DP	[16]	2017	Used Rényi's divergence to provide e-Rényi differential privacy with respect to alpha (a) called 'privacy order'. Instead of sensitivity, Rényi's divergence takes place. Evaluated Rényi divergence of Laplace, randomised, and Gaussian	Rényi divergence on Laplace, randomised, and Gaussian	(α, ε)-RDP	Used Rényi's divergence to replace sensitivity of DP	Privacy bound, randomised response

(Continued)

Table 9.2 (Continued) Comparison of variants of differential privacy on basis of functioning, noise addition mechanism, privacy criterion, key functioning, and evaluation type

Mechanism name	Ref #	Year	Major functioning	Noise addition distribution	Privacy criterion	Key functioning	Evaluation type
One-sided differential privacy	[33]	2017	Identified that whether data is sensitive or not and only added noise in the sensitive data during release	Laplace mechanism	(ε)-DP	Private data identification and answering count queries	MAE, MRE
Concentrated DP	[34]	2016	Proposed a zero and mean concentrated differential privacy mechanism that allows sharper analysis of many privacy-preserving criterions	Gaussian mechanism	(μ,τ)-CDP	Analysis over privacy criterions	Theoretical
Individual DP	[35]	2016	Proposed a notion of DP that allow data controllers to manage and calibrate the distortion at their end	Discrete Laplace	(ε)-iDP	Protected range queries w.r.t baseline	Error rate
Bayesian-DP	[15]	2015	First they studied influence of correlation on data privacy and adversary and afterward developed a perturbation mechanism for correlated data in a Bayesian way	Used Laplace in a Bayesian way	(z)-DP	Bayesian way of protecting privacy	Privacy leakage
Free Lunch-DP	[36]	2011	Studied the effect of background knowledge on differential privacy and formulised that it has strong effect	Modified Laplace	(ε)-freelunch	Background knowledge effect and deniability	Community edges

(Continued)

Table 9.2 (Continued) Comparison of variants of differential privacy on basis of functioning, noise addition mechanism, privacy criterion, key functioning, and evaluation type

Mechanism name	Ref #	Year	Major functioning	Noise addition distribution	Privacy criterion	Key functioning	Evaluation type
Random DP	[37]	2011	Randomly adding or subtracting a new observation in the query result will have small effect	Laplace	$(\theta, \gamma, \varepsilon)$-DP	Added randomly drawn noise as new observation	Loss
ZKP-DP	[38]	2011	Worked over a zero-knowledge-based privacy-preservation mechanism that gives results similar to differential privacy	Laplace	(ε)-DP	Comparison of ZKP with DP	–

Abbreviations: DP, differential privacy; MAE, mean absolute error; MRE, mean random error.

Moving to the next variant, a known relaxation mechanism of differential privacy was introduced by Mironov [16], which works on the phenomenon of Rényi divergence. The presented work can serve as an analytical tool to measure the privacy of any mechanism via ε-RényiDP method. Furthermore, instead of sensitivity, the authors worked Rényi's divergence and evaluated Rényi's divergence for Laplace, randomised, and Gaussian mechanisms. Keeping in view the basic functioning of differential privacy, the authors also computed privacy bounds for the developed tool. One more work that focuses on the identification of sensitive data in a data set and obfuscating only the sensitive data has been carried out by Doudalis et al. [33]. Authors first studied the problem of private data sharing in a case where only some specific elements or subsects in shared data are sensitive. So, instead of obfuscating the complete data set, the authors developed 'one-sided differential privacy'. The authors used a traditional Laplace mechanism to add up differentially private noise in the selected data reduced mean absolute and mean relative error. Extensive simulations over TIPPERs and DP-bench data set were performed in case of count queries.

A theoretical evaluation of a novel variant of differential privacy, 'Concentrated DP', was carried out by the authors in [34]. The proposed work discussed the privacy guarantee, extension, simplifications, and lower bounds of concentrated differential privacy strategy by providing an analysis over privacy criterions. The article proposed a zero and mean concentrated differential privacy mechanism that allows a sharper privacy analysis of certain privacy-preserving criterions. Moving to another private noise addition variant of differential privacy, a variant named 'Individual DP' was proposed by Soria-Comas et al. [35]. The authors stated that enforcing a strict differential privacy guarantee over a complete data set reduces its utility, and the majority of data gets affected because of small sensitive data. Therefore, the authors proposed a notion of individual differential privacy that allows data controllers to manage and calibrate the distortion of data at their end to provide more utility along with enhanced privacy. The proposed model uses the discrete Laplace mechanism to add Laplacian noise in protected range queries. The authors evaluated error rates and showed that the given strategy outperforms the baseline method.

Protecting correlated databases is a burden for privacy-preserving strategies, and sometimes it becomes quite difficult to protect these data sets because of a strong correlation with other databases. To overcome this issue, the authors in [15] proposed a Bayesian differential privacy-based privacy protection strategy for correlated data. At first, the authors studied the influence of the correlation on data privacy and adversary; afterward the authors developed a differentially private perturbation mechanism for correlated data in a Bayesian way. To add up further, the authors used a traditional Laplace mechanism but modified the noise addition criterion in a Bayesian manner and tested this method over artificial and social network data for privacy leakage. The results show that the proposed strategy outperforms the basic differential privacy mechanisms for specific databases that are correlated. Apart from correlation, sometimes differential privacy mechanism also suffer from a background knowledge attack, in which adversaries having certain background knowledge try to infer into private information of data owners. One of such background knowledge effect over differential privacy mechanisms was studied by Kifer et al. [36]. From a strong theoretical analysis, the authors formulised and showed that background knowledge has strong affect over privacy leakage, despite enforcing a strong privacy guarantee. Furthermore, the authors worked on contingency tables and evaluated community edges to study the effect.

A noise addition framework, 'random differential privacy' (RDP) was proposed by Hall et al. [37]. The authors claimed to propose a relaxed definition of differential privacy in which they added a new observation randomly in the output records. Complying to the basic definition of differential privacy, the authors theoretically evaluated that this will have minimal impact on the output data release. Moreover, the authors showed a composition property of a proposed strategy and compared it with baseline differential privacy to ensure the correctness. From an evaluation perspective, the authors released the RDP histogram data and showed that the utility loss via their proposed strategy outperforms that of baseline traditional differential privacy strategy. Another work in the field of integration of differential privacy with other privacy techniques was carried out by the authors in [38], in which the authors formulated a zero-knowledge-based differential privacy mechanism that gives similar results to that of actual differential privacy. Further analysis of the provided strategy showed that applying Laplace noise over zero-knowledge-based differential privacy is a viable solution to protect averages, fractions, histograms, and similar data and query types.

Keeping all of this in view, plenty of work is being carried out in the field of differential privacy. However, there is still room for new and modern mechanisms because of the increasing demand of differential privacy in nearly every sector, ranging from IoT to computing and from industry to academia.

9.3 LITERATURE REVIEW

Over the past few years, industries and researchers have started intensively working on decentralised distributed ledger technology (or blockchain) because of the boom of Bitcoin cryptocurrency. Researchers started understanding the paradigm of distributed transactions, and this led to the formation of the modern blockchain. The modern blockchain network allows users to effectively mine and store their transactional records in a distributed manner via a consensus algorithm over a decentralised ledger topology. The major reason behind the integration of blockchain is its secure nature that enhances trust of users in the network; therefore researchers are actively integrating blockchain in daily life scenarios. In this section, we provide a detailed overview of available literature from three different perspectives regarding possibilities of the integration of blockchain, edge computing, and smart grid and its privacy (Figure 9.1 and Table 9.3).

9.3.1 Blockchain work considering edge computing and its privacy

Blockchain as a network provides security and immutability, although, the common blockchain users such as mobile phones, smart meters, and healthcare IoT devices are not capable of performing extensive computing and mining functions. Therefore, there is always a need to add a new paradigm that enhances the computing capabilities of the network. For this reason, researchers integrated edge computing in this decentralised scenario and developed a framework that has the functionalities of both secure blockchain and computationally rich edge computing.

Despite this integration, a blockchain-based edge network is not completely private and has a large number of privacy risks. Therefore, taking necessary measures to protect the privacy of such systems is also needed. In this subsection, we discuss three major domains in which work has been carried from the perspective of blockchain, edge computing, and its privacy issues.

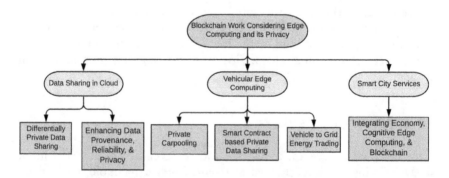

Figure 9.1 The blockchain approaches implemented in the domain of edge computing, and its privacy can be classified into data sharing in cloud, vehicular edge computing, and smart city services.

Table 9.3 Edge computing from its privacy perspective in domain of its basic working technique, enhancement, simulator, consensus mechanism, and blockchain type

Domain	Ref #	Year	Name of mechanism	Major technique	Enhancement	Simulator used	Consensus	Type of blockchain
Data sharing in cloud	[40]	2018	Differentially private data sharing in cloud with blockchain	Integrated differential privacy with blockchain to control privacy budget at user end	• Smart contract-based privacy budget control • Timeliness of data utility for federated cloud	Hyperledger	Byzantine	Private
	[41]	2017	ProvChain	A decentralised cloud storage for tamper-proof provenance data	• Provenance data security and privacy • Reliability and overhead	Apache Jmeter	Proof of provenance	Public
Vehicular edge computing	[42]	2018	Blockchain-based efficient and secure data sharing for vehicular edge networks	Reputation-based mechanism for data sharing using smart contracts in vehicles	• Data sharing • Security	–	Manual consensus	Consortium
	[43]	2018	Private carpooling for vehicular fog computing	Anonymous authentication, destination matching, and one-to-many matching is carried out using blockchain	• Enhanced one-to-many matching privacy • Computational cost and communication overhead	–	PoS	Private

(Continued)

Table 9.3 (Continued) Edge computing from its privacy perspective in domain of its basic working technique, enhancement, simulator, consensus mechanism, and blockchain type

Domain	Ref #	Year	Name of mechanism	Major technique	Enhancement	Simulator used	Consensus	Type of blockchain
	[44]	2018	Blockchain and edge enabled V2G energy trading	Stackelberg game-based resource allocation strategy for edge-based blockchain	• Block creation success probability • Network profit	–	PoW	Consortium
Smart city services	[45]	2019	Blockchain-based cognitive framework for IoT edge	Spatio-temporal smart contracts for IoT-based smart cities	• Processing and end-to-end delays • Mechanism processing time • Economy of smart cities	–	Proof of payment	Public

Abbreviations: IoT, Internet of Things; PoS, proof-of-stake; PoW, proof-of-work; V2G, vehicle to grid.

1. *Data Sharing in the Cloud:* Cloud computing is a well-researched domain for many years, and both industry and academia are getting the benefits from cloud computing in the most proliferate manner. The cloud is being used in many business models, and approximately every industry that involves real-time users is utilising the advantages of cloud computing. As discussed, the field of cloud computing has being explored, but the notion of decentralised cloud computing or blockchain-based cloud computing is a new paradigm and research is being carried out in this field. The actual issue is the integration of blockchain and the cloud, which raises questions: How to get the maximum benefits from the integration of cloud computing and decentralised blockchain technology? What is the best application scenario for their integration? What are the possible issues that needs to be tackled during this integration? One major concern among these issues is privacy among the decentralised cloud. As blockchain works on a transparent ledger phenomenon, the users' data has more privacy risk as compared to traditional cloud storage.

   To protect data during cloud data sharing, some work has been carried out to ensure privacy. For example, the authors in [40] presented a privacy-preservation strategy that effectively integrates differential privacy with a blockchain-based cloud network. The authors used the functionality of the smart contract to control the privacy budget of differential privacy and it enhances timeliness and data utility for data sharing in a decentralised manner. The authors claimed that differential privacy can efficiently protect cloud users' privacy if it is integrated with cloud computing, and in doing so, the authors evaluated their proposed strategy using Hyperledger platform and built a private blockchain to carry out data sharing. Another work that discusses the enhancement of privacy in a cloud-based blockchain is carried out by Liang et al. [41]. The authors provided a secure architecture for decentralised cloud storage for tamper-proof provenance data along with enabling transparency in the blockchain network. The presented work enhanced reliability and overhead of the network and implemented the architecture using Apache Jmeter platform. Furthermore, the authors claimed that their presented strategy consumes less computational power as compared to other similar works in the field of cloud computing and blockchain.

2. *Vehicular Edge Computing:* Vehicular edge computing is being used by certain autonomous and manually controlled vehicles to enhance user experience and to reduce latency during communication. For example, edge nodes are used as intermediaries to establish communication between vehicles and the cloud. Such architecture usually comprises three layers, the cloud, edge, and user layers. These layers perform their functions individually and form a complete network.

However, this network has security and privacy flaws; therefore, research is examining this integrated blockchain scenario to get the maximum benefit of it.

One such work was carried out by Kang et al. [42]; the authors proposed a consortium blockchain-based private data sharing mechanism for a vehicular edge network. The mechanism performs its operation on the basis of reputation, which is derived by a smart contract that controls data sharing and privacy. Furthermore, as shown in Table 9.4, the work ensured data sharing security and confidentiality via programmable smart contract. Similarly, the authors in [43] proposed a decentralised carpooling mechanism that operated on edge technology. Vehicles in the proposed mechanism are connected to road side units (RSUs) to report their real-time parking slot preference, availability, and requests. The provided scenario is at a risk of losing privacy; therefore, the authors integrated anonymous authentication, destination matching, and one-to-many matching algorithm to effectively protect such privacy. The authors used a proof-of-stake (PoS) consensus mechanism to mine a block in a private blockchain network. Furthermore, the authors ensured matching privacy protection along with reducing computational cost and communication overhead.

The domain of energy trading of electric vehicles (EVs) using blockchain and edge computing is explored by Zhou et al. [44]. The authors proposed a game-theoretic energy-trading strategy for a decentralised edge network by using the Stackelberg game model for resource allocation. The authors used a consortium blockchain for secure and private energy trading by restricting the activity of users to a certain level. Moreover, the presented work claimed to enhance block creation success probability along with improving overall network profit. The provided research highlighted the integration of blockchain and edge computing along with its privacy requirement for a vehicular network. The presented work efficiently protects security and privacy in their respective domains; however, there is still room available for research in this particular field.

3. *Smart City Services:* Rapid urbanisation in recent years has raised certain social, economic, and environmental problems. Researchers are focusing to solve these problems by introducing the concept of 'smart cities'. Smart cities are upgraded versions of traditional cities; however, smart cities will carry out the most efficient use of resources and will provide better services to residents to enhance quality of life. Similarly, the paradigm of the blockchain also has certain features that could be beneficial in enhancing these technologies of the smart grid to provide the most optimal services. For example, the trust-free nature, democracy, transparency, decentralisation, and security are some features that need to be considered during the development of smart cities, and blockchain technology provides these features in the

Table 9.4 Smart grid and edge computing with their basic working technique, enhancement, simulator, consensus mechanism, and blockchain type

Domain	Ref #	Year	Name of mechanism	Major technique	Enhancement	Simulator used	Consensus	Type of blockchain
Smart grid edge authentication	[46]	2019	PBEM-SGN	Smart-contract-based group signature and channel authorisation mechanism to ensure users validity	• Privacy protection • Energy security • Gas cost	Geth	–	Private
Vehicle to grid network	[47]	2019	V2G energy trading via integrating blockchain and edge	Contract-theory-based incentive mechanism for social welfare and resource allocation	• Maximise utility	–	Majority	Consortium
	[48]	2019	SURVIVOR	Location and utility function based framework for secure energy trading	• Latency reduction • Throughput	–	PoW	Public

(Continued)

Table 9.4 (Continued) Smart grid and edge computing with their basic working technique, enhancement, simulator, consensus mechanism, and blockchain type

Domain	Ref #	Year	Name of mechanism	Major technique	Enhancement	Simulator used	Consensus	Type of blockchain
Edge resource management	[49]	2018	Auction for edge resource allocation in mobile blockchain	Multilayer neural network for optimal auction approach	• Revenue • Winning probability	Tensor flow	PoW	Public
	[50]	2018	SWM auction in edge-based blockchain	Optimal resource allocation while taking allocative externality into account	• Social welfare • Truthfulness	Geth	PoW	Public

Abbreviation: PoW, proof-of-work.

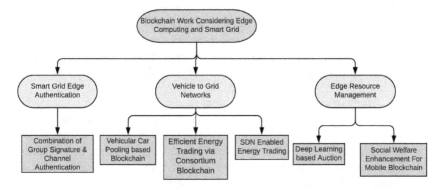

Figure 9.2 The blockchain approaches implemented in the domain of smart grid and edge computing can precisely be classified into smart grid edge authentication, vehicle to grid networks, and edge resource management.

best manner. Similarly, the integration of edge computing can facilitate this architecture with more computational power and low latency because of its tremendous features. Therefore, researchers are moving towards the integration of blockchain and edge computing in smart cities and its applications (Figure 9.2).

One such effort is carried out by Abdur Rahman et al. [45]; the authors proposed a blockchain-based cognitive framework for IoT edge of smart city. The proposed work uses spatio-temporal smart contract to enhance security and privacy in the decentralised network. Furthermore, the proposed strategy efficiently enhances end-to-end delays and processing time along with providing an economic solution to smart city applications. By viewing the discussion, it can be said that the smart city and its integration with blockchain and edge computing is a new paradigm; however, there is vast scope in this field and plenty of research still needs to be done.

9.3.2 Blockchain work considering edge computing and smart grid

As discussed previously, blockchain provides a secure and trustworthy platform to integrate multiple real-life scenarios, and integrating blockchain with edge computing and smart grid is one of them. The immutable nature of blockchain ensures validity, and similarly, the resource-rich environment of edge computing provides strong computational power. Therefore, researchers started integrating the merger of these two in certain other domains, and a smart grid is one of them. In the field of decentralised fog-enabled smart grid, research has been carried out in three major domains: Smart grid edge authentication, vehicle to grid (V2G) networks, and edge resource management. In this section, we provide a detailed overview of these three domains from technical perspectives.

1. *Smart Grid Edge Authentication:* Using edge computational capacity for smart grid operations has been discussed in literature, and plenty of work has been carried out. Similarly, increasing the security of such networks by integrating blockchain is also being considered as state-of-the-art research problem and certain work is being oriented in that direction [51]. One such work was carried out by Gai et al. [46] in context of edge-computing-based smart grid using permissioned blockchain. The authors developed a privacy-preserving mechanism based on a group signature and channel authorisation phenomenon to ensure validity of any communication in the network as shown in Table 9.5. Moreover, the authors claimed to enhance energy security and gas cost along with improving the privacy of the network. To evaluate the proposed strategy, the authors used the Geth simulator and developed a complete private blockchain with authentication, validation, and functioning nodes. This field of edge authentication using blockchain for smart grid is fairly new, and many challenges are yet to be solved. Therefore, research should be carried out in this field to overcome challenges.

2. *Vehicle to Grid Network:* Vehicular networks are among the fastest-growing networks right now because of the rapid increase in production of autonomous and modern electrical vehicles (EVs). This is leading towards the development of smart cities and smart transportation, in which vehicles will be connected with each other to carry out efficient information exchange and to take intelligent decisions such as traffic control, navigation, parking, carpooling, etc [60]. These connected EVs can be used for various purposes, for example, energy sustainability (by supplying surplus energy to grid), road safety (by reporting real-time congestion or accident information), crowd avoiding (enhancing parking availability), etc. Research is being carried out to develop modern solutions for smart transportation, and while doing so, researchers are integrating various recent technologies such as blockchain and edge computing with these vehicular networks.

One such work that efficiently integrates the vehicular fog computing network with blockchain to preserve privacy and to enhance efficiency of carpooling has been carried out by Li et al. [61]. The authors developed a private carpooling mechanism using conditionally anonymous authentication and private proximity matching to preserve user privacy. The mechanism was developed using Miracl platform in a private blockchain scenario. Despite that authors used a PoS consensus mechanism, the authors claimed that they efficiently enhanced and reduced computational cost and communication overhead. Another similar work that incorporates vehicle to grid energy trading via blockchain and edge

Table 9.5 Smart grid and its privacy requirement by focusing over their basic working technique, enhancement, simulator, consensus mechanism, and blockchain type

Domain	Ref #	Year	Name of mechanism	Major technique	Enhancement	Simulator used	Consensus	Type of blockchain
Smart grid monitoring and aggregation	[52]	2018	Grid monitoring	Trust-based system deployment for grid participants via smart contract	• Security • Trust	–	Manual	Public
	[53]	2018	Private aggregation on blockchain for power grid	Multiple-pseudonym-based privacy for private aggregation	• Privacy • Computational cost • Time complexity	–	–	Private
Vehicle to grid network	[54]	2018	Secure EV charging in energy blockchain and reputation-based DBFT consensus	Contract-theory-based resource allocation	• Utility • Optimal energy price & demand	–	DBFT	Private
	[55]	2017	Blockchain-based EV tariff decision	Automated charging station selection depending on distance and pricing	• Reliability • Transparency	–	PoW	Public
	[56]	2018	BBARS	Anonymous vehicle rewarding mechanism using dual public key cryptosystems	• Time cost	Miracl GMP	PoW	Public

(Continued)

Table 9.5 (Continued) Smart grid and its privacy requirement by focusing over their basic working technique, enhancement, simulator, consensus mechanism, and blockchain type

Domain	Ref #	Year	Name of mechanism	Major technique	Enhancement	Simulator used	Consensus	Type of blockchain
Miscellaneous	[57]	2018	Private tariff decisions via blockchain	Smart-contract-based tariff decision for blockchain	• Trust • Communication overhead	Solidity	PoW	Public
	[58]	2019	Keyless signature scheme for decentralised smart grid	Automated access control for smart grid via decentralised consensus	• Computational time • Scalability • Storage cost	Geth	PBFT	Public
Energy trading	[59]	2019	Private auction-based energy trading	Exploiting Laplace and exponential mechanism of differential privacy to protect auction privacy	• Prevented privacy leakage • Social welfare • Network benefit	Python	PoW	Consortium

Abbreviations: BBARS, blockchain-based anonymous rewarding scheme; DBFT, democratic byzantine fault tolerance; EV, electric vehicle; PBFT, practical byzantine fault tolerance; PoW, proof-of-work.

computing has been carried out by researchers in [47]. The proposed work maximised the utility of energy trading by using the contract-the-ory-based incentivisation mechanism. The proposed mechanism ensures non-negative social welfare along with efficient resource allocation to ensure users' trust and to increase the participation of users. Furthermore, the authors developed a majority consensus mechanism and deployed and tested this consensus over a consortium blockchain architecture.

Similarly, the authors in [48] developed a blockchain-based energy trading model for vehicle to grid networks in which they used edge as a service and also integrated software-defined networks to speed up the complete process. The proposed work uses location and utility function-based frameworks for secure energy trading and reducing latency in the network along with providing a high throughput. One difference from these mechanisms was that this method uses a public blockchain network in which every vehicle can join and play their part. Through simulation-based experiments, the authors claimed that the complete system is secure and effective in carrying out energy trading. After analysing the proposed work, it can be said that the field of EVs operating over edge computing, smart grid, and block-chain have a lot of potential that needs further exploration.

3. *Edge Resource Management:* Resource management has always been a vital domain from the perspective of scientists because of shortage or resources, whether they are spectrum resources, energy resources, or edge storage resources. However, resource management in the field of blockchain-based edge computing is a new paradigm and not enough work has been carried out in this domain. This field is still under exploration, and modern mechanisms for edge computing resources are being developed. Basically, blockchain requires high computa-tional power, and this demand can be filled using the computational-rich edge nodes. However, managing and allocating these edge nodes is the new challenge in that case.

To overcome this issue in context of mobile blockchain networks, the authors in [49] developed a deep learning-based approach to model and manage edge resource scarcity. The authors claimed that blockchain-based mobile applications and mobile networks are going to be one of the most prominent entities in future; however, these mobile networks can easily run out of computational capacity because of limited hardware. Therefore, edge computing can serve as a saviour in this domain, and edge com-puting-based blockchain for mobile edge networks can serve as saviour. The authors developed a multilayer neural network auction approach to allocate mobile nodes a desirable edge computing resource accordingly. The proposed mechanism enhances the revenue of the network and main-tains an adequate winning probability. The authors tested this mecha-nism over the TensorFlow library by using PoW consensus over a public blockchain and ensured that their mechanism can be applied to practical edge-based decentralised mobile networks.

The work in context of social welfare maximisation for edge-computing-based mobile blockchain is carried out by Jiao et al. [50]. The authors developed an optimal auction mechanism while taking into account allocative externality. The proposed strategy was tested using the Geth simulation tool over a public blockchain scenario, and the authors claimed that their proposed work increases users' trust by providing them a secure platform to carry out decentralised resource auction. Although, there is some work that is available in the field of mobile edge-computing-based blockchain networks, this research field has a lot of potential and could be explored further to develop more efficient applications (Figure 9.3).

9.3.3 Blockchain work considering smart grid and its privacy

Using blockchain as a commodity to enhance smart grid performance is a new paradigm, and plenty of research is developing efficient blockchain-based solutions to enhance the performance of the smart grid. For example, some are moving towards the generation, and transmission of decentralised energy, and others are doing work in storage and management. In this section, we discuss certain scenarios in which blockchain has been integrated and the fruitful results that have been obtained.

1. *Smart Grid Monitoring and Aggregation:* Reporting real-time data to grid utility is one of the major requirements of modern-day energy systems. For this purpose, smart meters are being used; a smart meter is an electric device that enables bidirectional communication

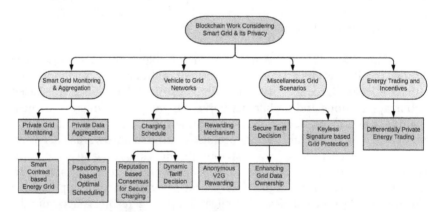

Figure 9.3 The blockchain approaches implemented in the domain of smart grid and its privacy can precisely be classified into smart grid monitoring and aggregation, vehicle to grid networks, miscellaneous grid scenarios, and energy trading and incentives.

between the grid distributer utility and the meter itself [61]. This bidirectional communication enables smart meters to perform certain tasks such as real-time reading collection, demand response management, etc. As discussed previously, one of the most prominent features of this smart metering network is real-time reporting and monitoring of power. This information is viable and is sometimes necessary to carry out load forecasting. Furthermore, with the increasing number of RERs, reporting generation and usage of RERs is also becoming an important parameter in this reporting. This real-time reporting is necessary; however, it comes with certain security and privacy issues. For example, if this data gets leaked to some third-party intruder, then the intruder can easily get to know the lifestyle of residents. Similarly, a security flaw in this can cause a breakdown in the complete system, and this situation can even lead to blackouts.

To overcome these issues, researchers are integrating blockchain with smart grid, and, it is providing good results. Blockchain being a decentralised distributed ledger provides a secure and trusted platform to carry out real-time reporting. One such work was carried out by the authors in [52]. The authors developed a decentralised grid monitoring strategy using a smart contract. The proposed strategy uses a manual consensus mechanism over public blockchain and ensures that the strategy enhances trust and security in the network by providing real-time immutable reporting. Similarly, Guan et al. [53] proposed a decentralised privacy preservation mechanism for in-group privacy preservation using blockchain technology. The proposed work used multiple-pseudonym-based privacy for private aggregation of smart metering data, and the authors ensured that the smart meter users could trust this network from context of privacy. Furthermore, the proposed strategy enhances the computational cost along with improving time complexity. For this purpose, the authors evaluated private blockchain concept and presented a complete simulation-based model on the concept of private blockchain-based grid monitoring.

2. *Vehicle to Grid Network:* RERs hold a great importance in the development of modern energy systems and have great potential to solve energy issues of the modern world. Nowadays, the concept of distributed RERs is being implemented, which will consist of decentralised distributed microgrids operating on their own for their required energy generation. Similarly, chargeable EVs are also being integrated with these distributed microgrids to enhance demand response of the network. Because these EVs are capable of charging and discharging their energy at the time of need, they can be used to overcome energy crisis in an emergency situation. Therefore, research is being carried out in the integration of EVs and the smart grid. Although, this integration is fruitful in many ways, it brings certain technical challenges as well such as security and privacy issues. To overcome these

challenges, blockchain emerged as one of the most viable options, in which these technologies can be integrated by ensuring security in their communication.

One such work in the field of EV charging and discharging in a blockchain network has been carried out by Su et al. [54]. The authors developed a reputation distributed byzantine fault tolerant consensus mechanism and integrated this consensus mechanism with contract-theory-based resource allocation to ensure security in the network. Furthermore, the simulation results showed that the work provides optimal energy pricing for distributed EV discharging along with enhancing overall utility of the network. Similarly, the authors in [55] developed a privacy-preserving mechanism to make dynamic tariff decisions of EVs within a public blockchain network. Furthermore, the authors worked on the phenomenon of automated charging station selection depending on distance and pricing; the algorithm suggests to EVs their most suitable charging station along with keeping their data confidential. In this way, the trust of EVs in the network can be enhanced, and reliability can be maximised. The work on anonymous vehicle rewarding mechanisms was carried out by the authors in [56]. The developed mechanism, 'BBARS', uses a dual public key cryptography system to ensure security. The authors developed the complete framework over Miracl, PBC, and GMP platforms using PoW consensus on a public blockchain. Furthermore, the authors compared the time cost of their proposed mechanism and showed from simulation results that their work effectively enhances time cost to an optimal level.

3. *Energy Trading:* Modern smart homes are integrating RERs that are capable of producing a certain amount of electricity for their usage. However, this generated energy can be used in two ways: One for their personal use and other to store and sell. This concept of buying and selling surplus electricity produced at smart homes is known as microgrid energy trading [62]. Usually energy trading is carried out via developing certain auction strategies; however, traditional auction strategies lack trust because of their insecure nature. For example, users are worried about unfair decisions, along with leakage of their bids, etc. For this reason, scientists are developing more secure and trustworthy platforms. During the quest to do so, blockchain is being integrated with modern microgrids and energy-trading nodes to provide a secure and efficient platform. Integrating blockchain in this scenario provided fruitful results; however, in some specific scenarios, such as auction games, blockchain lacks privacy because of its transparent nature. Right now, researchers are trying to find more private solutions to carry out microgrid auctions along with securing users' privacy.

One such work that protects privacy of smart homes energy-trading auction has been carried out by the authors in [59]. The authors proposed a differentially private auction protection mechanism working on a decentralised consortium blockchain scenario. The aim of the proposed approach was to provide a secure and trustworthy atmosphere to microgrids in which they can trade their energy without the fear of losing their private data. To do so, the authors used Laplacian noise addition and exponential perturbation selection phenomenon of differential privacy. This dual protection mechanism ensures that no adversary can intrude into the private data of grid users despite being available over blockchain ledger. The proposed work also enhanced the utility, revenue, and network benefits of participating nodes along with providing a trustworthy platform for trading.

4. *Miscellaneous:* Along with all these works, some work carried out in the field of blockchain and smart grid can just not be linked with one specific field. Therefore, we discuss the work in this section. First of all, a work over private tariff decisions has been carried out by Knirsch et al. [57]. The authors evaluated the concept of smart-contract-based public blockchain to take intelligent tariff decisions. The authors developed a mechanism that creates a balance between energy generation and utilisation by efficiently managing energy prices according to usage habits of customers. Furthermore, the authors claimed to enhance the communication overhead of the network along with improving customers' trust. Another work to enhance security of the network by using keyless signature was carried out by the authors in [58]. The authors developed a smart automated access control mechanism for smart grid by via decentralised consensus. The proposed method was evaluated over Geth and simulation results showed that the strategy efficiently enhanced scalability, computational time, and storage cost of signatures and records.

9.4 DIFFERENTIALLY PRIVATE RER MONITORING FOR EDGE-COMPUTING-BASED SMART GRID OPERATING OVER BLOCKCHAIN

Real-time on-demand reporting of smart meter data from an edge-based aggregator can leak users' privacy, and any adversary can easily get to know private information regarding the usage, generation, and occupancy of smart homes. Furthermore, grid utilities can also use this data for certain financial and commercial purposes (e.g., increasing the price of electricity at the time of unavailability of RERs, etc.). Keeping this discussion in view, we develop a blockchain-based differentially private on-demand edge reporting (BDOR) strategy that protects privacy of smart meter users working on the phenomenon of decentralised edge-based nodes. In this section, we discuss the system model, design goal, adversary model, and proposed algorithm of BDOR strategy.

9.4.1 System model

The BDOR strategy consists of three major components: Smart home, fog aggregators, and smart database. Smart homes are entities that consist of smart meters, RERs, and home appliances. Smart meters are capable of real-time reporting of energy usage and generation. Similarly, RERs generate intermittent energy using the prospective source, such as solar, wind, biomass, etc. Furthermore, home appliances consume the energy irrespective whether its from grid or RERs. The second entity in DPOR model is a fog aggregator; the fog aggregator serves as an intermediary to control the information flow for a specific suburb or a small town. The aggregators collect data from smart meters and manage certain operations, such as billing, demand response, etc. Furthermore, these fog aggregators do also transmit on-demand readings to grid utility data-bases in the time of need. The last entity in DPOR is smart grid utilities; these utilities are large entities that have control and functional information of a large area (e.g., a city). They consist of control centres to control the ingoing and outgoing electricity; furthermore, they also contain large cloud databases for storage of information of smart meters and fog aggregators. These three components serve as a blockchain node at their prescribed level. For example, smart homes serve as a blockchain node and have the potential to equal to all others; however, the functionalities of smart home blockchain node is less compared to a fog aggregator. Therefore, the smart home has less computational capacity. Similarly, fog aggregators are linked with each other using a blockchain decentralised network and carry out dissemination accordingly. These fog aggregators also respond to on-demand reading request of grid utility and transmit differentially private data accordingly.

Furthermore, the grid utilities receive information from fog aggregators and store them in their cloud storage; afterward, this data is mined into blockchain via a consensus algorithm. The complete functioning diagrams of DPOR strategy is given in Figures 9.4 and 9.5.

In traditional smart grid on-demand reporting systems, data from smart meters is transmitted to fog aggregators, and these fog aggregators store this data to perform certain operations, such as calculating bills, enhancing demand response, etc. Furthermore, fog aggregators also respond to the request of grid utilities in case intermittent values of generation or utilisation at a specific time slot are needed. This reporting model is functional; however, it has two major flaws. One flaw is that from smart meter to fog aggregator, the data is transmitted without any specific security measure, and users' data is always prone to some attack, which can cause a change in the actual values. Therefore, a specific secure strategy or architecture is required that can enhance security to ensure users trust. Second, requesting on-demand reading from fog aggregators can result in a privacy leak because of intermittent

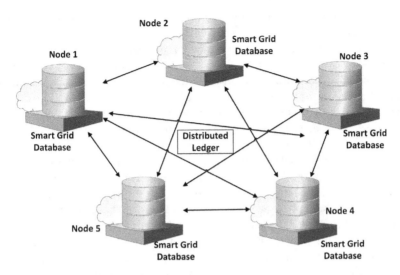

Figure 9.4 Differentially private on-demand reporting of fog aggregators based over blockchain by considering smart homes as blockchain nodes and a fog aggregator as a special blockchain node collecting all data from smart homes and regulating the behaviour of these smart homes with respect to instructions from smart grid utility database.

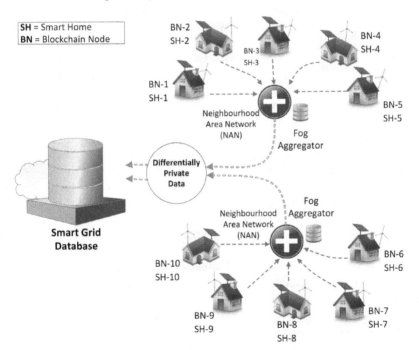

Figure 9.5 Application scenario of decentralised distributed smart grid databases linked with each other with a blockchain network.

reporting of intermittent values. Therefore, the integration of some privacy-preserving strategy is also required. Our proposed BDOR strategy efficiently overcomes both flaws by integrating secure blockchain and private differential privacy-preservation strategy in the traditional smart grid on-demand reporting scenario. In Figure 9.4, a group of smart homes mounted with smart meter and RERs can be seen; these smart homes are linked with fog aggregators. This collectively forms a neighbourhood area network (NAN). Smart homes, being a blockchain node, transmits its information to fog aggregator, which is responsible for aggregation in NAN. This forms a cluster type structure in which all smart homes are linked with each other and they unite to form a complete functional grid network. Smart homes are not directly linked with grid database; however, the stored values of these smart homes are transmitted to grid database via an intermediary fog aggregator, which is responsible for protecting privacy. In our scenario, we took the aggregator as trusted entity, and the grid database as untrusted adversary because of the nature of the job they are performing. Fog aggregator transmits differentially private protected reading to the grid database in case of on-demand request; moreover, after a specific interval of time, the aggregator transmits all stored or accumulated values to the grid database while maintaining privacy.

This NAN can also be termed a private blockchain network, in which each smart home node can only write or transmit the data to its respective for aggregator; however, the right of mining the block in a blockchain is restricted only to fog nodes. Furthermore, the fog aggregators compete with each other to solve a complex cryptographic puzzle to carry out PoW consensus among them. After receiving values from smart homes after a certain interval of time, every aggregator node tries to carry out consensus as soon as possible by competing with other aggregator nodes. The node mining the block first gets a reward accordingly; similarly, malicious nodes get removed in case of false reporting or mining.

Similarly, Figure 9.5 demonstrates the network of all grid databases that collect information from fog aggregators and store this data to cloud servers after carrying out consensus. The grid databases can ask for on-demand reading from fog aggregators to carry out future load forecasting at a specific time slot. Moreover, these databases collect monthly readings from fog aggregators to keep records of usage and generation and usage of energy. These databases compete with each other during consensus mechanism and solve PoW puzzles to mine a block; similarly, the winning database gets a reward.

The goal of the BDOR strategy is to provide a complete architecture in which each smart home can report its real-time data without the fear of data tampering or privacy loss. The complete algorithm of our proposed BDOR strategy is given in Algorithm 1. Moreover, the brief system model of our proposed BDOR strategy is given in Figures 9.4 and 9.5.

9.4.2 Design goals of BDOR strategy

In literature, research has been done over smart grid operating on blockchain and smart grid operating on edge computing. However, none of the work considered private on-demand reporting for smart grid operating on blockchain-based edge computing. Design goals of our BDOR strategy are as follows:

- Transmitting and recording complete real-time usage and generation data from smart meter to edge aggregators.
- Integrating private blockchain technology with aggregators acting as authoritative edge nodes and with smart meters functioning as blockchain nodes.
- Achieving security during transmission of data from smart meters to aggregators and from aggregators to grid utility.
- Preserving privacy during on-demand reading extraction of grid utility from edge aggregator nodes.
- Protecting instantaneous leakage of generation information of RERs in a smart home.
- Achieving differentially private on-demand reading transmission to ensure smart home privacy.

9.4.3 BDOR strategy adversary model

Edge-node-based decentralised aggregators have complete usage and generation data profiles of smart meters, and they are also receiving data from smart meters after every 10 minutes. Similarly, the grid utility can ask for data from these aggregators at any instant of time, however revealing this instantaneous usage and generation data raises certain privacy concerns. We assume that grid utility database is a curious-but-honest adversary.

**Algorithm 9.1   BDOR Algorithm for
Decentralised On-Demand RER Reporting**

Input: S(Users' Set), C_I(Consumption Info), G_I(Generation Info), T_s(Time Slot), U_P(Unit Price), U_{ID}(User ID), ε(Privacy Criterion), Δq(Sensitivity), X(Cycle Time)

Output: M_{UR}(Monthly Usage Reading), M_{GR}(Monthly Generation Reading), I_R(Instantaneous Reading), MAE(Mean Absolute Error), MRE(Mean Relative Error), B_I(Billing Info), Q_R(Query Result)

FUNCTION → Metering-Aggregation(S, X, C_I, G_I, I_s, U_{ID}, M_R, M_{GR})
1: **for** (i ← 1 to X) **do**
2:     **for** (j ← 1 to S) **do**
3:         Calculate $M_{UR} = M_{UR} + C_I$// Manage Demand Response
4:         Calculate $M_{GR} = M_{GR} + G_I$

5:    end for
6: end for
        //After a cycle time, e.g., 31 days
7: $BI = MUR * UP$
        //Transmitting this value for monthly billing
FUNCTION → ON-Demand- Reporting(S, C_I, G_I, I_R) //Grid Utility asks for on-demand reading (query)
8: for $i \leftarrow 1$ to N do
9:    $q(S, U_{ID}) \leftarrow S(U_{ID})$
10:   $\Delta q \leftarrow S_X$

11:   $P_r\left(\mathbb{F}(q, S, U_{ID}) = U_{ID}\right) \leftarrow \dfrac{\exp\left(\dfrac{\varepsilon.q\left(S, U_{ID}\right)}{2\Delta q}\right)}{\displaystyle\sum_{U_{ID}' \in S(U_{ID})} \exp\left(\dfrac{\varepsilon.q\left(S, U_{ID}'\right)}{2\Delta q}\right)}$

12: end for
13: $U_{ID} \leftarrow F(q, S, U_{ID})$ // Complete Probability Distribution
14: $q_R \leftarrow$ Query result selected from $U_{ID} \leftarrow F(q, S, U_{ID})$
15: $O_R \leftarrow$ Original answer
16: $E_r(MAE, MRE) \leftarrow$ error & accuracy parameter w.r.t O_R and Q_R
17: return $B_I, q_R, E_r(MAE, MRE)$

---

which can critically analyse users' data but will not alter this data in any way. Therefore, in our BDOR privacy-preserving strategy, we protect this private on-demand transmitting of data by perturbing it via exponential mechanism of differential privacy. To provide complete privacy, we use an exponential mechanism along with allowing query evaluation to grid adversary. From the collected on-demand reading data, the grid adversary cannot predict with confidence the absence or presence of any specific RER generation or presence or absence of residents in a smart home.

9.4.4 Proposed algorithm and flow chart

In the BDOR strategy, we preserve on-demand reporting privacy of smart meter users along with accumulating and aggregating the complete data of smart meters in decentralised edge nodes. To do so, we developed an algorithm, presented in Algorithm 9.1. To protect privacy of smart metering data, we use query exponential mechanism of differential privacy, in which the grid utilities can ask queries but will always get a perturbed differentially private output [63]. Moving towards the proposed algorithm, we divide the complete working of algorithm into two major functions: 'Metering Aggregation' and 'On-Demand Reporting'. First, in the Metering Aggregation function, real-time data from smart meter is collected from all meters within a specific range of aggregator, let's say a specific cluster size. Afterward, we calculate monthly usage reading and monthly generation reading by accumulating the real-time collected values. In the meantime, the aggregator also calculates certain side tasks such as calculating demand response, etc. However, a detailed discussion about these functionalities is out of the scope of this chapter. The same function does also calculate monthly or fortnightly bills using the unit price. Certain dynamic pricing algorithms can also be integrated in this step because the aggregator is a trusted entity and can be used for dynamic pricing.

The second function is the algorithm, On-Demand Reporting; this function is the core part of the BDOR strategy because of its dynamic differentially private nature. In this function, the grid utility is given the privilege of asking queries, such as 'How many houses are generating energy more than 500 w at a specific time slot?' or 'What is the maximum amount of energy being generated by a specific house right now?' and many other similar queries. These types of queries are highly private, and if the utility becomes an adversary, it can leak critical private information. Therefore, we use exponential mechanism of differential privacy during this query evaluation to protect users' privacy. In this reporting function, we first allowed the utility to ask queries, and when one utility ask queries from the aggregator database, then all required data related to a specific query is gathered autonomously. Afterward, an exponential differentially private functions makes a differentially private distribution for all input values within a range of a specific private parameter and a sensitivity value Δq. The formula used to formulise the distribution is as follows:

$$P_r\left(F\left(q, S, U_{ID}\right) = U_{ID}\right) \propto \frac{\exp\left(\frac{\varepsilon.q\left(S, U_{ID}\right)}{2\Delta q}\right)}{\sum_{U_{ID'} \in S(U_{ID})} \exp\left(\frac{\varepsilon.q\left(S, U'_{ID}\right)}{2\Delta q}\right)} \tag{9.3}$$

Once the distribution is finalised, the answer of query is selected exponentially from this distribution. This answer is further transmitted to the grid utility and is also stored in the aggregator to carry out error estimation analysis as described in the algorithm.

9.5 PERFORMANCE EVALUATION

To carry out performance evaluation of our BDOR privacy-preserving query mechanism, we consider two important parameters of differential privacy, mean absolute error (MAE) and mean relative error (MRE). These two parameters demonstrate the error in the values being reported via aggregators, so that the utility of the BDOR mechanism can be analysed. We carried out experiments over Python 3.0 by using its numerical libraries such as Pandas v0.24, NumPy v1.14. Furthermore, we conducted experiments over developed data of 1,000 smart homes and analysed query evaluation via exponential mechanism. Moreover, we performed experiments at different epsilon (ε) values such as 0.001, 0.01, 0.1, 0.3, 0.5, 0.7, and 0.7 to analyse all possible results. Furthermore, we carried out 1,000 iterations for every experiment to eradicate any numerical error. The detailed discussion about MAE and MRE and their evaluation at different privacy parameters is provided in the following section.

9.5.1 Mean absolute and mean relative error

Enhancing the utility of query evaluation by reducing the error rate is the actual requirement of any privacy-preserving query evaluation mechanism. To evaluate our BDOR approach, we consider MAE and MRE. Furthermore, to calculate MAE and MRE in a query evaluation scenario, we took a numeric query and its truthful answer and afterward we compared both numeric values. For example, the query 'what is the maximum value in a specific slot in which highest number of microgrids are their maximum energy'? has a specific numeric answer that can vary according to data set, we termed this specific numeric answer a 'truthful answer'. Once the aggregator receives such query, it forms a differentially private database using an exponential mechanism of differential privacy, and afterward a random answer according to its probability distribution is picked from that database.

1. *Mean Absolute Error:* MAE can be termed as the difference between two variables that are independent from each other. If A and B are two variables collected as a result of some observations, then MAE can be given as follows:

$$MAE = \frac{\sum_{y=1}^{n} |A_y - B_y|}{n} \tag{9.4}$$

In our scenario, let's say truthful value is the maximum number of users contribution in that slot, and the errored value is the picked answer from differentially private formulated distribution. Absolute error demonstrates the actual difference between the original and reported value; therefore, it can be termed a 'parameter' to evaluate the utility of privacy mechanism. For example, the maximum error rate it is, the lower is the utility.

Simulation graphs demonstrating MAE at different values of epsilon is presented in Figure 9.6. The data in the graph is collected after running the same experiment for 1,000 times and calculating the mean

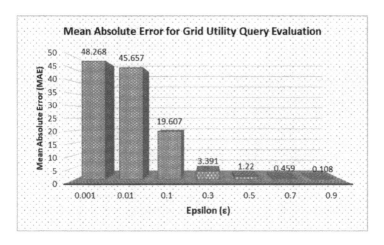

Figure 9.6 Mean absolute error for grid query evaluation at different epsilon values.

of results. The first bar in Figure 9.6 presents MAE value at epsilon value of 0.001; it can be seen from the graph that error value is maximum (48.268), which means maximum utility. When we move down the graph to epsilon value 0.01, the error value reduces a bit to 45.657, which still shows that significant level of privacy is maintained by the aggregator. Moving further to the next bar at epsilon value of 0.1, we see a sudden drop in error rate from 45.657 to 19.607. This happens because of the reduction in privacy. As privacy or indistinguishability of the result reduces, it causes a reduction in error value, which means more utility but less privacy. This trend keeps on going to the next bars, and a significant reduction can be observed. The value of error rate is minimum at epsilon value of 0.9, which is closest to the complete utility and no privacy. An error rate of 0.108 shows that after 1,000 careful observations, only mean 0.108 error is found in all reported readings, which demonstrates that almost every value was reported truthfully. This value of epsilon can be adjusted after careful observation and demonstration between grid utility and smart meter users.

2. *Mean Relative Error:* Another parameter that we considered while evaluating privacy and accuracy of BDOR strategy is MRE. Relative error refers to an error is which relative to its true value; for example, the ratio of deviation with respect to its true value can be termed a relative error. The formula for MRE is as follows:

$$MRE = \frac{\sum_{y=1}^{n} \left(\frac{A_y - B_y}{A_y} \right)}{n} \tag{9.5}$$

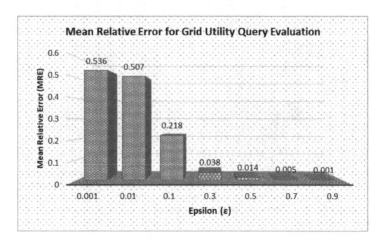

Figure 9.7 Mean relative error for grid query evaluation at different epsilon values.

The major difference in the formula of MRE and MAE is the ratio taken with respect to its relativeness from truthful value or truthful query answer. From experimental results in Figure 9.7, it can be observed that MRE value decreases with the increase in epsilon value. The value of MRE is maximum at the minimum epsilon (say 0.001), which ensures maximum privacy protection. Moving further to bigger values of epsilon, this value decreased guaranteeing more utility but less privacy. The value almost reduces to zero at epsilon value of 0.9, which ensures highest utility value. *After analysing all experimental results from MAE and MRE perspectives and relating them with the utility and privacy ratio, our BDOR mechanism efficiently protects the privacy of smart meter users during grid query evaluation.*

9.6 CHALLENGES AND FUTURE RESEARCH DIRECTIONS

Despite being a secure and decentralised platform, blockchain and its integration with edge computing is still vulnerable to plenty of research challenges that require major considerations. In this section, we highlight some critical domains and challenges to pave paths for future research.

9.6.1 Privacy issues

Even though the major focus of this chapter is proposing a privacy-preservation strategy for application oriented blockchain-based edge computing, privacy still remains one of the major challenges in the field of blockchain and its major domains because of its transparent and decentralised nature [28]. Blockchain users can never be comfortable in using blockchain-based applications, unless they are completely sure about their privacy protection. Currently, in a blockchain network, users are not even anonymous; they are pseudonymous. As the transactions, addresses, names, and so on are linked with each other in a pseudonym manner and are publicly available to every node in the ledger [64]. Therefore, after careful analysis of stored data, one can infer into private information and can easily track all activities. Furthermore, certain attacks such as linking attacks, correlation attacks, and background-knowledge attacks can be carried out on this data to infiltrate users' privacy. To date, many privacy-preserving strategies have been integrated in blockchain-based systems, and researchers have discussed the many advantages of by integrating various strategies. For example, the authors in [65] discussed the uses of integration of differential privacy in a blockchain scenario. Similarly, the authors in [66] worked on integration of k-anonymity with a blockchain scenario. Moreover, integration of blockchain with edge computing on one hand enhances its efficiency, but on the other hand, it also makes it vulnerable because edge nodes are more prone to such attacks because they carry a huge amount of network data, and in blockchain-based edge computing that has a copy of the ledger, it will provide a hacker all the private data of that network. Moreover, some researchers proposed an off-chain storage solution to protect this data, but this in turn is against the decentralised nature of blockchain. Currently, research is being carried out to protect the privacy of blockchain, but despite this, blockchain privacy is still vulnerable to attacks. Therefore, research needs to be carried out in this field to provide completely private systems.

9.6.2 Security issues

The security core of blockchain works on a cryptographic protection, which uses the phenomenon of public and private keys to protect the security of users. This concept is also used for transaction authentication that ensures the correct person gets the transactional amount. Despite this secure mechanism, the security of blockchain is still vulnerable because of certain attacks such as fault-injection attacks [67]. Furthermore, 51% attacks can also be one of the major security concerns of new blockchain system. Nevertheless, launching this 51% attack is near to impossible in large cryptosystems such as Bitcoin because of the high computational power involved. But still it is a big threat for newly launching blockchain applications because there system can easily be collapsed in case of such attack.

Furthermore, outsourcing the work of blockchain to edge and cloud nodes also raises plenty of security risks. For example, the authors in [51] outsourced the hectic computationally expensive work of blockchain to edge nodes, which on one hand is fruitful, but on the other hand, the complete network becomes more vulnerable [4]. Keeping all this in view, it can be said that blockchain-based edge computing is a secure integration, but it still has certain flaws that require proper addressing.

9.6.3 Consensus mechanism

Consensus can be a method to ensure mutual agreement of all nodes on a single point of interest. All consensus mechanisms consists of two basic steps; one is *block proposal* and the second is *decision factor*. Not all proposals meet consensus because of no mutual agreement among the nodes. Similarly, the concept of forking also causes certain legal transactions to get overruled by others, and thus, the chain gets disconnected for those specific transactions [1].

Similarly, the concept of intelligent incentivisation and punishment has also came up as a challenge for modern blockchain-based edge systems because of increasing vulnerabilities. In certain multiuser oriented scenarios, such as EV energy trading, where multiple parties are involved depending on their contribution (e.g., buyer, seller, charging station, grid utility, etc.), intelligent incentivisation scenarios are required that can distribute the complete generated profit according to the level of contribution. Furthermore, to avoid double spending, and similar attacks, proper penalties and punishment mechanisms should be designed depending on the need of application. Researchers are proposing numerous consensus algorithms that can overcome such issues; however, the development of a completely perfect application-oriented consensus mechanism is still a bit challenge that needs to be resolved.

9.6.4 Energy efficiency and throughput

The concept of energy efficiency and throughput of blockchain can be linked together because they behave in parallel to each other. For example, if a consensus mechanism is energy efficient, then it requires less computational power, and it will produce more blocks as compared to a mechanism that requires more computational power to mine a block because more powerful mining nodes will be required [68]. However, on the other hand, if throughput is very high, then it can be a serious reason for forking, in which two blocks can be formed at the same time, leaving one to get forked in the long run. Moving back to energy efficiency, some small blockchain nodes such as IoT devices cannot always meet the minimum requirement of mining because of their limited computational complexity; however, in these systems, large mining giants get control over the system, which raises the issue of centralisation. Therefore, research needs to be carried out to develop an energy-efficient and throughput-maximising consensus mechanism while keeping in view the other mentioned issues.

9.6.5 Machine learning

Almost all modern applications involve the use of machine learning to make it more efficient. There is no doubt that machine learning algorithms helps authorities to handle large amount of data and to get useful information from that data. Furthermore, machine learning is also being used to detect vulnerabilities and anomalies in a network, and these algorithms can even predict a failure before it occurs [69]. Some work is being carried out by researchers to integrate machine learning and deep learning with blockchain and edge computing in various scenarios such as developing access control policy [70], enhancing blockchain security [71], overcoming deanonymisation attacks [72], and many more.

However, these machine learning and deep learning algorithms are not mature enough and require more detailed development. Therefore, research is required to be carried out to enhance this field and to get maximum benefits from deep learning in a decentralised blockchain scenario.

9.7 CONCLUSION

Blockchain is an emerging paradigm that attracted attention of researchers and scientists because of its decentralised nature. Nowadays, blockchain is being integrated with plenty of applications to enhance their security. Cryptographic key encryption technology used in blockchain ensures that data does not get leaked to any third-party adversary. However, adding or mining a new block in a blockchain network is computationally expensive and requires specific computational resources. Therefore, integrating it in real-life scenarios and applications such as smart grid, healthcare, and so on is still a challenge. To overcome this challenge, edge computing was devised as a solution. Resource-rich nodes of edge computing provide an efficient platform that can help blockchain nodes to overcome this resource scarcity. One such example is edge-computing-based smart grid operating over a decentralised blockchain platform. In this chapter, we propose a blockchain-based differentially private on-demand edge reporting (BDOR) mechanism that is based on blockchain-based edge computing, and further uses differential privacy to efficiently protect the privacy of smart meter users. We carried out extensive experiments to study differential privacy and its integration with this scenario from the perspective of MAE and MRE. Performance evaluation of the proposed strategy shows that it efficiently protects smart meter users' data privacy along with providing a secure decentralised blockchain-based atmosphere for its users.

REFERENCES

1. M. Belotti, N. Božić, G. Pujolle, and S. Secci, 'A vademecum on blockchain technologies: When, which and how,' *IEEE Communications Surveys & Tutorials*, vol. 21, no. 4, pp. 3796–3838, 2019.
2. M. S. Ali, M. Vecchio, M. Pincheira, K. Dolui, F. Antonelli, and M. H. Rehmani, 'Applications of blockchains in the Internet of Things: A comprehensive survey,' *IEEE Communications Surveys & Tutorials*, vol. 21, no. 2, pp. 1676–1717, 2018.

3. P. Porambage, J. Okwuibe, M. Liyanage, M. Ylianttila, and T. Taleb, 'Survey on multi-access edge computing for Internet of Things realization,' *IEEE Communications Surveys & Tutorials*, vol. 20, no. 4, pp. 2961–2991, 2018.

4. R. Yang, F. R. Yu, P. Si, Z. Yang, and Y. Zhang, 'Integrated blockchain and edge computing systems: A survey, some research issues and challenges,' *IEEE Communications Surveys & Tutorials*, vol. 21, no. 2, pp. 1508–1532, 2019.

5. A. Dorri, S. S. Kanhere, and R. Jurdak, 'Blockchain in Internet of Things: Challenges and solutions,' *arXiv preprint arXiv:1608.05187*, 2016.

6. Y. Zhang, Z. Hao, and S. Wang, 'A differential privacy support vector machine classifier based on dual variable perturbation,' *IEEE Access*, vol. 7, pp. 98 238–98 251, 2019.

7. L. Sweeney, 'k-anonymity: A model for protecting privacy,' *International Journal of Uncertainty, Fuzziness and Knowledge-Based Systems*, vol. 10, no. 5, pp. 557–570, 2002.

8. A. Machanavajjhala, J. Gehrke, D. Kifer, and M. Venkitasubramaniam, 'l-diversity: Privacy beyond k-anonymity,' in *22nd International Conference on Data Engineering (ICDE'06)*. IEEE, 2006, pp. 24–24.

9. N. Li, T. Li, and S. Venkatasubramanian, 't-closeness: Privacy beyond k-anonymity and l-diversity,' in *IEEE 23rd International Conference on Data Engineering*, 2007, pp. 106–115.

10. Y. Li and H. Shen, 'Equi-width data swapping for private data publication,' in *International Conference on Parallel and Distributed Computing, Applications and Technologies*. IEEE, 2009, pp. 231–238.

11. D. Kifer, 'On estimating the swapping rate for categorical data,' in *Proceedings of the 21th ACM SIGKDD International Conference on Knowledge Discovery and Data Mining*, Sydney, Australia, 2015, pp. 557–566.

12. A. T. Hasan, Q. Jiang, J. Luo, C. Li, and L. Chen, 'An effective value swapping method for privacy preserving data publishing,' *Security and Communication Networks*, vol. 9, no. 16, pp. 3219–3228, 2016.

13. Gentry, Craig. 'Fully homomorphic encryption using ideal lattices,' in *Proceedings of the forty-first annual ACM symposium on Theory of computing*, pp. 169–178, 2009.

14. K.-A. Shim, 'A survey of public-key cryptographic primitives in wireless sensor networks,' *IEEE Communications Surveys & Tutorials*, vol. 18, no. 1, pp. 577–601, 2015.

15. B. Yang, I. Sato, and H. Nakagawa, 'Bayesian differential privacy on correlated data,' in *Proceedings of the ACM SIGMOD International Conference on Management of Data*, 2015, pp. 747–762.

16. I. Mironov, 'Rényi differential privacy,' in *IEEE 30th Computer Security Foundations Symposium (CSF)*, 2017, pp. 263–275.

17. M. Maeda and Y. Unagami, 'History information anonymization method and history information anonymization device for anonymizing history information,' Jul. 3 2018, US Patent 10,013,576.

18. D. Goldschlag, M. Reed, and P. Syverson, 'Onion routing for anonymous and private internet connections,' Naval Research Lab Washington DC Center For High Assurance Computing Systems..., Technical Reports, 1999.

19. V. Estivill-Castro and L. Brankovic, 'Data swapping: Balancing privacy against precision in mining for logic rules,' in *International Conference on Data Warehousing and Knowledge Discovery*. Springer, Florence, Italy, 1999, pp. 389–398.

20. C. Dwork, 'Differential privacy,' *Encyclopedia of Cryptography and Security*, pp. 338–340, 2011.
21. Z. Fan, P. Kulkarni, S. Gormus, C. Efthymiou, G. Kalogridis, M. Sooriyabandara, Z. Zhu, S. Lambotharan, and W. H. Chin, 'Smart grid communications: Overview of research challenges, solutions, and standardization activities,' *IEEE Communications Surveys & Tutorials*, vol. 15, no. 1, pp. 21–38, 2012.
22. M. R. Asghar, G. Dán, D. Miorandi, and I. Chlamtac, 'Smart meter data privacy: A survey,' *IEEE Communications Surveys & Tutorials*, vol. 19, no. 4, pp. 2820–2835, 2017.
23. U. Greveler, B. Justus, and D. Loehr, 'Multimedia content identification through smart meter power usage profiles,' *Computers, Privacy and Data Protection*, vol. 1, no. 10, 2012.
24. M. Mukherjee, L. Shu, and D. Wang, 'Survey of fog computing: Fundamental, network applications, and research challenges,' *IEEE Communications Surveys & Tutorials*, vol. 20, no. 3, pp. 1826–1857, 2018.
25. Y. Mao, C. You, J. Zhang, K. Huang, and K. B. Letaief, 'A survey on mobile edge computing: The communication perspective,' *IEEE Communications Surveys & Tutorials*, vol. 19, no. 4, pp. 2322–2358, 2017.
26. S. Nakamoto et al., 'Bitcoin: A peer-to-peer electronic cash system,' 2008. https://bitcoin.org/bitcoin.pdf.
27. M. U. Hassan, M. H. Rehmani, and J. Chen, 'Differential privacy techniques for cyber physical systems: A survey,' *IEEE Communications Surveys & Tutorials, in Print*, 2019.
28. M. U. Hassan, M. H. Rehmani, and J. Chen, 'Privacy preservation in blockchain based IoT systems: Integration issues, prospects, challenges, and future research directions,' *Future Generation Computer Systems*, vol. 97, pp. 512–529, 2019.
29. T. Zhu, G. Li, W. Zhou, and S. Y. Philip, 'Differentially private data publishing and analysis: A survey,' *IEEE Transactions on Knowledge and Data Engineering*, vol. 29, no. 8, pp. 1619–1638, 2017.
30. Z. Ji, Z. C. Lipton, and C. Elkan, 'Differential privacy and machine learning: a survey and review,' *arXiv preprint arXiv:1412.7584*, 2014.
31. A. Cheu, A. Smith, J. Ullman, D. Zeber, and M. Zhilyaev, 'Distributed differential privacy via shuffling,' in *Annual International Conference on the Theory and Applications of Cryptographic Techniques*. Springer, Darmstadt, Germany, 2019, pp. 375–403.
32. N. Fernandes, M. Dras, and A. McIver, 'Generalised differential privacy for text document processing,' in *International Conference on Principles of Security and Trust*. Springer, Prague, Czech Republic, 2019, pp. 123–148.
33. S. Doudalis, I. Kotsogiannis, S. Haney, A. Machanavajjhala, and S. Mehrotra, 'One-sided differential privacy,' *arXiv preprint arXiv:1712.05888*, 2017.
34. M. Bun and T. Steinke, 'Concentrated differential privacy: Simplifications, extensions, and lower bounds,' in *Theory of Cryptography Conference*. Springer, Beijing, China, 2016, pp. 635–658.
35. J. Soria-Comas, J. Domingo-Ferrer, D. Sánchez, and D. Megías, 'Individual differential privacy: A utility-preserving formulation of differential privacy guarantees,' *IEEE Transactions on Information Forensics and Security*, vol. 12, no. 6, pp. 1418–1429, 2017.

36. D. Kifer and A. Machanavajjhala, 'No free lunch in data privacy,' in *Proceedings of the 2011 ACM SIGMOD International Conference on Management of data*. ACM, Athens, Greece, 2011, pp. 193–204.

37. R. Hall, A. Rinaldo, and L. Wasserman, 'Random differential privacy,' *arXiv preprint arXiv:1112.2680*, 2011.

38. J. Gehrke, E. Lui, and R. Pass, 'Towards privacy for social networks: A zero-knowledge based definition of privacy,' in *Theory of Cryptography Conference*. Springer, Providence, RI, 2011, pp. 432–449.

39. D. Desfontaines and B. Pejó, 'Sok: Differential privacies,' *arXiv preprint arXiv:1906.01337*, 2019.

40. M. Yang, A. Margheri, R. Hu, and V. Sassone, 'Differentially private data sharing in a cloud federation with blockchain,' *IEEE Cloud Computing*, vol. 5, no. 6, pp. 69–79, 2018.

41. X. Liang, S. Shetty, D. Tosh, C. Kamhoua, K. Kwiat, and L. Njilla, 'Provchain: A blockchain-based data provenance architecture in cloud environment with enhanced privacy and availability,' in *Proceedings of the 17th International Symposium on Cluster, Cloud and Grid Computing*. IEEE Press, 2017, pp. 468–477.

42. J. Kang, R. Yu, X. Huang, M. Wu, S. Maharjan, S. Xie, and Y. Zhang, 'Blockchain for secure and efficient data sharing in vehicular edge computing and networks,' *IEEE Internet of Things Journal*, vol. 6, no. 3, pp. 4660–4670, 2018.

43. M. Li, L. Zhu, and X. Lin, 'Efficient and privacy-preserving carpooling using blockchain-assisted vehicular fog computing,' *IEEE Internet of Things Journal*, vol. 6, no. 3, pp. 4573–4584, 2018.

44. Z. Zhou, L. Tan, and G. Xu, 'Blockchain and edge computing based vehicle-to-grid energy trading in energy Internet,' in *2nd IEEE Conference on Energy Internet and Energy System Integration (EI2)*, 2018, pp. 1–5.

45. M. A. Rahman, M. M. Rashid, M. S. Hossain, E. Hassanain, M. F. Alhamid, and M. Guizani, 'Blockchain and IoT-based cognitive edge framework for sharing economy services in a smart city,' *IEEE Access*, vol. 7, pp. 18 611–18 621, 2019.

46. K. Gai, Y. Wu, L. Zhu, L. Xu, and Y. Zhang, 'Permissioned blockchain and edge computing empowered privacy-preserving smart grid networks,' *IEEE Internet of Things Journal*, vol. 6, no. 5, pp. 7992–8004, 2019.

47. Z. Zhou, B. Wang, M. Dong, and K. Ota, 'Secure and efficient vehicle-to-grid energy trading in cyber physical systems: Integration of blockchain and edge computing,' *IEEE Transactions on Systems, Man, and Cybernetics: Systems*, 2019.

48. A. Jindal, G. S. Aujla, and N. Kumar, 'Survivor: A blockchain based edge-as-a-service framework for secure energy trading in SDN-enabled vehicle-to-grid environment,' *Computer Networks*, vol. 153, pp. 36–48, 2019.

49. N. C. Luong, Z. Xiong, P. Wang, and D. Niyato, 'Optimal auction for edge computing resource management in mobile blockchain networks: A deep learning approach,' in *IEEE International Conference on Communications (ICC)*. IEEE, 2018, pp. 1–6.

50. Y. Jiao, P. Wang, D. Niyato, and Z. Xiong, 'Social welfare maximization auction in edge computing resource allocation for mobile blockchain,' in *IEEE International Conference on Communications (ICC)*. IEEE, 2018, pp. 1–6.

51. M. A. Khan and K. Salah, 'Iot security: Review, blockchain solutions, and open challenges,' *Future Generation Computer Systems*, vol. 82, pp. 395–411, 2018.

52. J. Gao, K. O. Asamoah, E. B. Sifah, A. Smahi, Q. Xia, H. Xia, X. Zhang, and G. Dong, 'Gridmonitoring: Secured sovereign blockchain based monitoring on smart grid,' *IEEE Access*, vol. 6, pp. 9917–9925, 2018.

53. Z. Guan, G. Si, X. Zhang, L. Wu, N. Guizani, X. Du, and Y. Ma, 'Privacy-preserving and efficient aggregation based on blockchain for power grid communications in smart communities,' *IEEE Communications Magazine*, vol. 56, no. 7, pp. 82–88, 2018.

54. Z. Su, Y. Wang, Q. Xu, M. Fei, Y.-C. Tian, and N. Zhang, 'A secure charging scheme for electric vehicles with smart communities in energy blockchain,' *IEEE Internet of Things Journal*, in Print, 2018.

55. F. Knirsch, A. Unterweger, and D. Engel, 'Privacy-preserving blockchain-based electric vehicle charging with dynamic tariff decisions,' *Computer Science-Research and Development*, vol. 33, no. 1–2, pp. 71–79, 2018.

56. H. Wang, Q. Wang, D. He, Q. Li, and Z. Liu, 'Bbars: Blockchain-based anonymous rewarding scheme for v2g networks,' *IEEE Internet of Things Journal*, vol. 6, no. 2, pp. 3676–3687, 2019.

57. F. Knirsch, A. Unterweger, G. Eibl, and D. Engel, 'Privacy-preserving smart grid tariff decisions with blockchain-based smart contracts,' in *Sustainable Cloud and Energy Services*. Springer, 2018, pp. 85–116.

58. H. Zhang, J. Wang, and Y. Ding, 'Blockchain-based decentralized and secure keyless signature scheme for smart grid,' *Energy*, vol. 180, pp. 955–967, 2019.

59. M. U. Hassan, M. H. Rehmani, and J. Chen, 'Deal: Differentially private auction for blockchain based microgrids energy trading,' *IEEE Transactions on Services Computing*, in Print, 2019.

60. F. Cunha, L. Villas, A. Boukerche, G. Maia, A. Viana, R. A. Mini, and A. A. Loureiro, 'Data communication in vanets: Protocols, applications and challenges,' *Ad Hoc Networks*, vol. 44, pp. 90–103, 2016.

61. S. Finster and I. Baumgart, 'Privacy-aware smart metering: A survey,' *IEEE Communications Surveys & Tutorials*, vol. 16, no. 3, pp. 1732–1745, 2014.

62. M. U. Hassan, M. H. Rehmani, R. Kotagiri, J. Zhang, and J. Chen, 'Differential privacy for renewable energy resources based smart metering,' *Journal of Parallel and Distributed Computing*, vol. 131, pp. 69–80, 2019.

63. T. Zhu, G. Li, W. Zhou, and S. Y. Philip, 'Preliminary of differential privacy,' in *Differential Privacy and Applications*. Springer, 2017, pp. 7–16.

64. J. Xie, H. Tang, T. Huang, F. R. Yu, R. Xie, J. Liu, and Y. Liu, 'A survey of blockchain technology applied to smart cities: Research issues and challenges,' *IEEE Communications Surveys & Tutorials*, 2019.

65. M. U. Hassan, M. H. Rehmani, and J. Chen, 'Differential privacy in blockchain technology: A futuristic approach,' *arXiv preprint arXiv:1910.04316*, 2019.

66. J. Wang, M. Li, Y. He, H. Li, K. Xiao, and C. Wang, 'A blockchain based privacy-preserving incentive mechanism in crowdsensing applications,' *IEEE Access*, vol. 6, pp. 17 545–17 556, 2018.

67. O. Boireau, 'Securing the blockchain against hackers,' *Network Security*, vol. 2018, no. 1, pp. 8–11, 2018.

68. M. A. Ferrag, M. Derdour, M. Mukherjee, A. Derhab, L. Maglaras, and H. Janicke, 'Blockchain technologies for the internet of things: Research issues and challenges,' *IEEE Internet of Things Journal*, vol. 6, no. 2, pp. 2188–2204, 2018.

69. M. Du, F. Li, G. Zheng, and V. Srikumar, 'Deeplog: Anomaly detection and diagnosis from system logs through deep learning,' in *Proceedings of the 2017 ACM SIGSAC Conference on Computer and Communications Security*. ACM, New York, 2017, pp. 1285–1298.

70. A. Outchakoucht, E. Hamza, and J. P. Leroy, 'Dynamic access control policy based on blockchain and machine learning for the internet of things,' *International Journal of Advanced Computer Science and Applications*, vol. 8, no. 7, pp. 417–424, 2017.

71. D. Somdip, 'A proof of work: Securing majority-attack in blockchain using machine learning and algorithmic game theory,' PhD dissertation, Modern Education and Computer Science Press, 2018.

72. M. A. Harlev, H. Sun Yin, K. C. Langenheldt, R. Mukkamala, and R. Vatrapu, 'Breaking bad: De-anonymising entity types on the bitcoin blockchain using supervised machine learning,' in *Proceedings of the 51st Hawaii International Conference on System Sciences*, 2018.

Index

Note: Page numbers in bold and italics refer to tables and figures, respectively.